The University of Chicago School Mathematics Project

Skills Link

Cumulative Practice Sets
Student Book

The McGraw·Hill Companies

Photo Credits

Cover—©Gregory Adams/Getty Images, cover, *top right*; Getty Images, cover, *center*; ©Tony Hamblin: Frank Lane Picture Agency/CORBIS, cover, *bottom left*

Photo Collage—Herman Adler Design

---

www.WrightGroup.com

Printed in the United States of America.

Send all inquiries to:
Wright Group/McGraw-Hill
P.O. Box 812960
Chicago, IL 60681

ISBN 978-0-07-622504-0
MHID 0-07-622504-6

3 4 5 6 7 8 9 MAL 15 14 13 12 11 10

The McGraw·Hill Companies

Contents

Practice Sets Correlated to Grade 4 Goals

Content	*Everyday Mathematics* Grade 4 Grade-Level Goals	Grade 4 Practice Sets
Number and Numeration		
Place value and notation	**Goal 1.** Read and write whole numbers up to 1,000,000,000 and decimals through thousandths; identify places in such numbers and the values of the digits in those places; translate between whole numbers and decimals represented in words and in base-10 notation.	2, 8, 11, 12, 13, 14, 20, 26, 28, 33, 35, 37, 38, 39, 42, 43, 48, 57, 59, 63, 69, 70, 73, 77, 81, 89
Meanings and uses of fractions	**Goal 2.** Read, write, and model fractions; solve problems involving fractional parts of a region or a collection; describe and explain strategies used; given a fractional part of a region or a collection, identify the unit whole.	17, 18, 21, 22, 25, 32, 40, 48, 51, 53, 54, 58, 63, 67, 69, 72, 73, 79, 82, 85, 89, 91, 92
Number theory	**Goal 3.** Find multiples of whole numbers less than 10; find whole number factors of numbers.	15, 25, 66, 90
Equivalent names for whole numbers	**Goal 4.** Use numerical expressions involving one or more of the basic four arithmetic operations and grouping symbols to give equivalent names for whole numbers.	2, 10, 14, 31, 38, 89, 90
Equivalent names for fractions, decimals, and percents	**Goal 5.** Use numerical expressions to find and represent equivalent names for fractions and decimals; use and explain a multiplication rule to find equivalent fractions; rename fourths, fifths, tenths, and hundredths as decimals and percents.	27, 32, 55, 56, 60, 67, 68, 69, 70, 79, 85, 86, 87, 90
Comparing and ordering numbers	**Goal 6.** Compare and order whole numbers up to 1,000,000,000 and decimals through thousandths; compare and order integers between –100 and 0; use area models, benchmark fractions, and analyses of numerators and denominators to compare and order fractions.	3, 28, 29, 32, 43, 44, 53, 57, 60, 62, 72, 79, 83, 86, 88, 89, 92
Operations and Computation		
Addition and subtraction facts	**Goal 1.** Demonstrate automaticity with basic addition and subtraction facts and fact extensions.	1, 4, 5, 26, 29, 35, 36, 37, 38, 42, 47, 59, 69, 84
Addition and subtraction procedures	**Goal 2.** Use manipulatives, mental arithmetic, paper-and-pencil algorithms, and calculators to solve problems involving the addition and subtraction of whole numbers and decimals through hundredths; describe the strategies used and explain how they work.	2, 4, 6, 11, 12, 15, 17, 19, 23, 24, 26, 27, 28, 29, 30, 31, 33, 34, 37, 40, 41, 42, 43, 44, 46, 47, 50, 51, 54, 55, 57, 59, 69, 70, 71, 72, 77, 80, 84, 87, 89, 92
Multiplication and division facts	**Goal 3.** Demonstrate automaticity with multiplication facts through $10 * 10$ and proficiency with related division facts; use basic facts to compute fact extensions such as $30 * 60$.	2, 4, 5, 8, 10, 11, 14, 15, 17, 19, 20, 21, 22, 25, 26, 33, 35, 36, 37, 38, 39, 40, 43, 45, 46, 47, 48, 50, 54, 55, 57, 59, 62, 63, 64, 65, 66, 67, 71, 74, 75, 81, 84, 85, 86, 88, 90, 91, 92, 93
Multiplication and division procedures	**Goal 4.** Use mental arithmetic, paper-and-pencil algorithms, and calculators to solve problems involving the multiplication of multidigit whole numbers by 2-digit whole numbers and the division of multidigit whole numbers by 1-digit whole numbers; describe the strategies used and explain how they work.	2, 6, 7, 12, 19, 21, 24, 27, 28, 29, 33, 36, 37, 38, 39, 40, 41, 43, 45, 46, 47, 48, 50, 54, 64, 66, 70, 73, 74, 81, 82, 84, 87, 89, 92, 93
Procedures for addition and subtraction of fractions	**Goal 5.** Use manipulatives, mental arithmetic, and calculators to solve problems involving the addition and subtraction of fractions with like and unlike denominators; describe the strategies used.	54, 56, 65, 74

Computational estimation	**Goal 6.** Make reasonable estimates for whole number and decimal addition and subtraction problems, and whole number multiplication and division problems; explain how the estimates were obtained.	9, 15, 23, 29, 37, 38, 39, 43, 48, 52, 66, 85, 93
Models for the operations	**Goal 7.** Use repeated addition, skip counting, arrays, area, and scaling to model multiplication and division.	14, 28, 46, 50, 51, 60
Data and Chance		
Data collection and representation	**Goal 1.** Collect and organize data or use given data to create charts, tables, bar graphs, line plots, and line graphs.	21, 29, 36, 41
Data analysis	**Goal 2.** Use the maximum, minimum, range, median, mode, and graphs to ask and answer questions, draw conclusions, and make predictions.	13, 14, 16, 24, 40, 53, 72, 75, 79
Qualitative probability	**Goal 3.** Describe events using *certain, very likely, likely, unlikely, very unlikely, impossible* and other basic probability terms; use *more likely, equally likely, same chance, 50–50, less likely,* and other basic probability terms to compare events; explain the choice of language.	52, 59, 76, 92
Quantitative probability	**Goal 4.** Predict the outcomes of experiments and test the predictions using manipulatives; summarize the results and use them to predict future events; express the probability of an event as a fraction.	18, 36, 52, 59, 76
Measurement and Reference Frames		
Length, weight, and angles	**Goal 1.** Estimate length with and without tools; measure length to the nearest $\frac{1}{4}$ inch and $\frac{1}{2}$ centimeter; estimate the size of angles without tools.	16, 30, 33, 35, 49, 50, 74, 75, 77, 86
Area, perimeter, volume, and capacity	**Goal 2.** Describe and use strategies to measure the perimeter and area of polygons, to estimate the area of irregular shapes, and to find the volume of rectangular prisms.	8, 60, 61, 62, 63, 64, 65, 66, 70, 77, 84, 85, 90
Units and systems of measurement	**Goal 3.** Describe relationships among U.S. customary units of length and among metric units of length.	23, 34, 35, 38, 60, 69, 76, 78, 81, 86
Coordinate systems	**Goal 4.** Use ordered pairs of numbers to name, locate, and plot points in the first quadrant of a coordinate grid.	9, 55, 58, 65
Geometry		
Lines and angles	**Goal 1.** Identify, draw, and describe points, intersecting and parallel line segments and lines, rays, and right, acute, and obtuse angles.	2, 3, 49, 50, 77, 82
Plane and solid figures	**Goal 2.** Describe, compare, and classify plane and solid figures, including polygons, circles, spheres, cylinders, rectangular prisms, cones, cubes, and pyramids, using appropriate geometric terms including *vertex, base, face, edge,* and *congruent.*	4, 5, 6, 7, 9, 45, 70, 82, 83
Transformations and symmetry	**Goal 3.** Identify, describe, and sketch examples of reflections; identify and describe examples of translations and rotations.	76, 77, 78, 81, 83
Patterns, Functions, and Algebra		
Patterns and functions	**Goal 1.** Extend, describe, and create numeric patterns; describe rules for patterns and use them to solve problems; use words and symbols to describe and write rules for functions that involve the four basic arithmetic operations and use those rules to solve problems.	2, 4, 7, 9, 13, 14, 16, 17, 18, 25, 28, 29, 30, 31, 35, 48, 50, 51, 55, 56, 60, 63, 68, 70, 80, 88, 89, 93
Algebraic notation and solving number sentences	**Goal 2.** Use conventional notation to write expressions and number sentences using the four basic arithmetic operations; determine whether number sentences are true or false; solve open sentences and explain the solutions; write expressions and number sentences to model number stories.	5, 24, 25, 26, 27, 32, 33, 38, 39, 45, 51, 52, 58, 64, 69, 73, 77, 78, 89, 92
Order of operations	**Goal 3.** Evaluate numeric expressions containing grouping symbols; insert grouping symbols to make number sentences true.	4, 5, 13, 26, 27, 35, 41, 63, 78, 84
Properties of the arithmetic operations	**Goal 4.** Apply the Distributive Property of Multiplication over Addition to the partial-products multiplication algorithm.	39

Name ______________________ Date __________ Time __________

Grade 3 Review: Number and Numeration

1. Write the decimal value for the shaded part of the grid.

____

2. Shade the grid to show the decimal.

0.67

3. Anya has 20 marbles.

- $\frac{1}{5}$ of them are blue. Write a **B** in $\frac{1}{5}$ of the marbles.
- $\frac{1}{4}$ of them are red. Write an **R** in $\frac{1}{4}$ of the marbles.
- $\frac{1}{2}$ of them are yellow. Write a **Y** in $\frac{1}{2}$ of the marbles.
- The rest are green. How many of the marbles are green? __________
- What fraction of the marbles are green? __________

4. Write the multiples of 5.

5, 10, ______, ______, ______, ______, ______

Name Date Time

Grade 3 Review: Number and Numeration

5. Write 10 more names for 21 in the box.

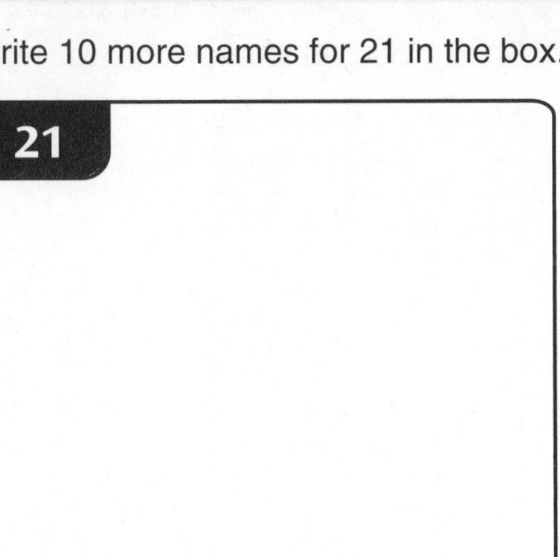

6. Circle the greater number in each pair.

a. 7,964
7,694

b. 35,014
35,140

c. 861,285
816,285

d. 604,003
604,103

---

7. Shade $\frac{1}{4}$ of each figure. Write the equivalent fraction.

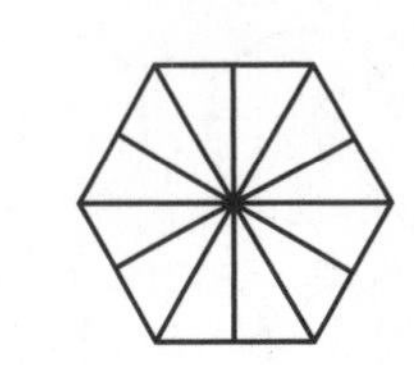

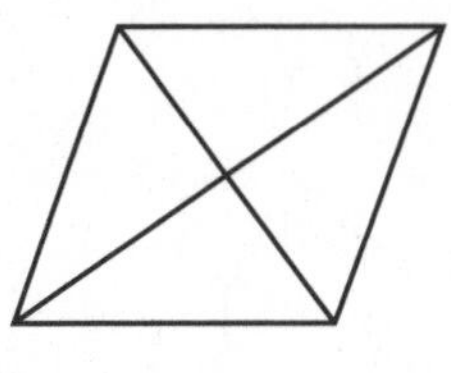

______ ______ ______

8. The chart shows the amounts raised by children in each grade.

Use <, > or = to write a number sentence that compares the amounts of money raised by Grade 3 and Grade 5.

____

Amounts Raised

Grade	Amount (in dollars)
2	$1,248
3	$1,264
4	$1,285
5	$1,259

9. What is the total number of blocks?

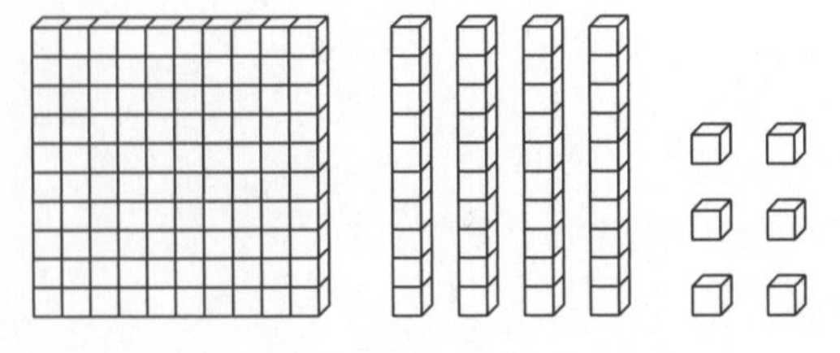

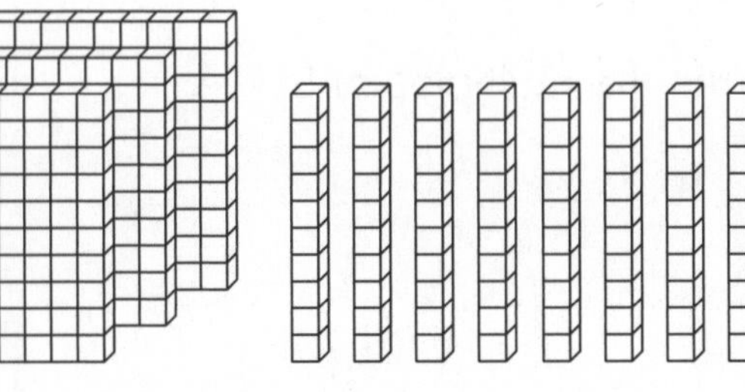

____

Name _______________ Date _______________ Time _______________

Grade 3 Review: Operations and Computation

Solve.

1. 60,000 + 30,000 = __________

2. 40,000 + 20,000 = __________

3. 70,000 − 50,000 = __________

4. 900,000 − 800,000 = __________

Write the fact families.

5. 54, 6, 9

_____ × _____ = _____

_____ × _____ = _____

_____ ÷ _____ = _____

_____ ÷ _____ = _____

6. 7, 4, 28

_____ × _____ = _____

_____ × _____ = _____

_____ ÷ _____ = _____

_____ ÷ _____ = _____

Multiply.

7. $\begin{array}{r} 61 \\ \times\ 8 \\ \hline \end{array}$

8. $\begin{array}{r} 427 \\ \times\ 5 \\ \hline \end{array}$

9. $\begin{array}{r} 56 \\ \times\ 9 \\ \hline \end{array}$

10. $\begin{array}{r} 360 \\ \times\ 7 \\ \hline \end{array}$

Make a ballpark estimate. Write a number model to show your estimate.

11. 796 − 387

_____ − _____ = _____

12. 403 − 105

_____ − _____ = _____

13. Five children each bring 3 cans of food for the food drive. Draw a picture and write a number sentence to show how many cans of food the children brought all together.

Name Date Time

Grade 3 Review: Data and Chance

Use the animal pictures to complete Problem 1.

1. Complete the tally chart.

Number of Legs on Animals	
Number of Legs	**Number of Animals**
0	
2	
4	
6	
8	

2. Use the tally chart to complete the graph.

Number of Legs on Animals

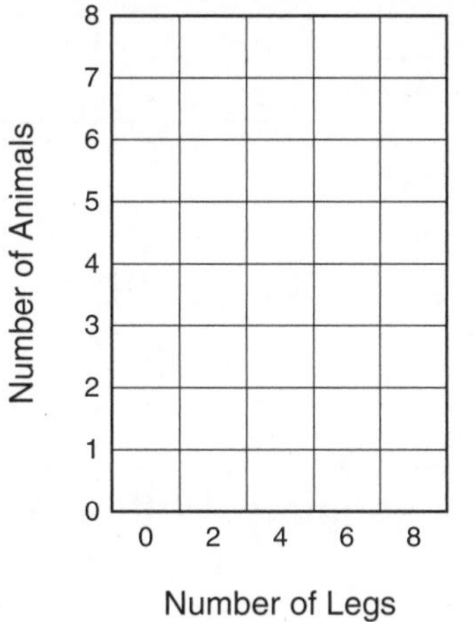

Use the graph to answer Problems 3–8.

3. What is the maximum of the data set? __________

4. What is the minimum of the data set? __________

5. What is the range of the data set? __________

6. What is the median of the data set? __________

7. What is the mean of the data set? __________

8. What is the mode of the data set? __________

Name Date Time

Grade 3 Review: Data and Chance

Look at the bag of shapes. How likely are you to pick the given shape out of the bag without looking? Write *certain, very likely, likely, unlikely, very unlikely,* or *impossible.*

9. a circle ____________

10. any shape ____________

11. a hexagon ____________

12. a triangle or square ____________

13. a pentagon ____________

14. a triangle ____________

Look at the bag of shapes above. Write the answers.

15. What are the chances of picking a circle out of the bag without looking? ________ out of ________

16. What are the chances of picking *any* shape out of the bag without looking? ________ out of ________

17. What are the chances of picking a triangle or a square out of the bag without looking? ________ out of ________

18. What are the chances of picking a pentagon out of the bag without looking? ________ out of ________

19. What are the chances of picking a shape that is *not* a square out of the bag without looking? ________ out of ________

20. What are the chances of picking a hexagon, a circle, or a square out of the bag without looking? ________ out of ________

Name ______________________ Date ____________ Time ____________

Grade 3 Review: Measurement and Reference Frames

Tell whether each angle shows a $\frac{1}{4}$, $\frac{1}{2}$, $\frac{3}{4}$, or full turn.

1. __________

2. __________

3. __________

4. Draw a figure with a perimeter of 24 units.

What is the **area** of your figure?

__________ square units

5. Draw a figure with an area of 28 square units.

What is the **perimeter** of your figure?

__________ units

6. Katie is 47 inches tall. How many feet tall is she? __________

How many inches are left over? __________

7. A regulation football field is 100 yards long.
How many feet long is a regulation football field? __________

8. Draw the hour and minute hands to show 3:38 P.M.

What time will it be 39 minutes later?

______ : ______

9. Write the time in the afternoon shown on the clock. Circle A.M. or P.M.

______ : ______ A.M. P.M.

What time was it 57 minutes earlier?

Circle A.M. or P.M. ______ : ______ A.M. P.M.

Name Date Time

Grade 3 Review: Geometry

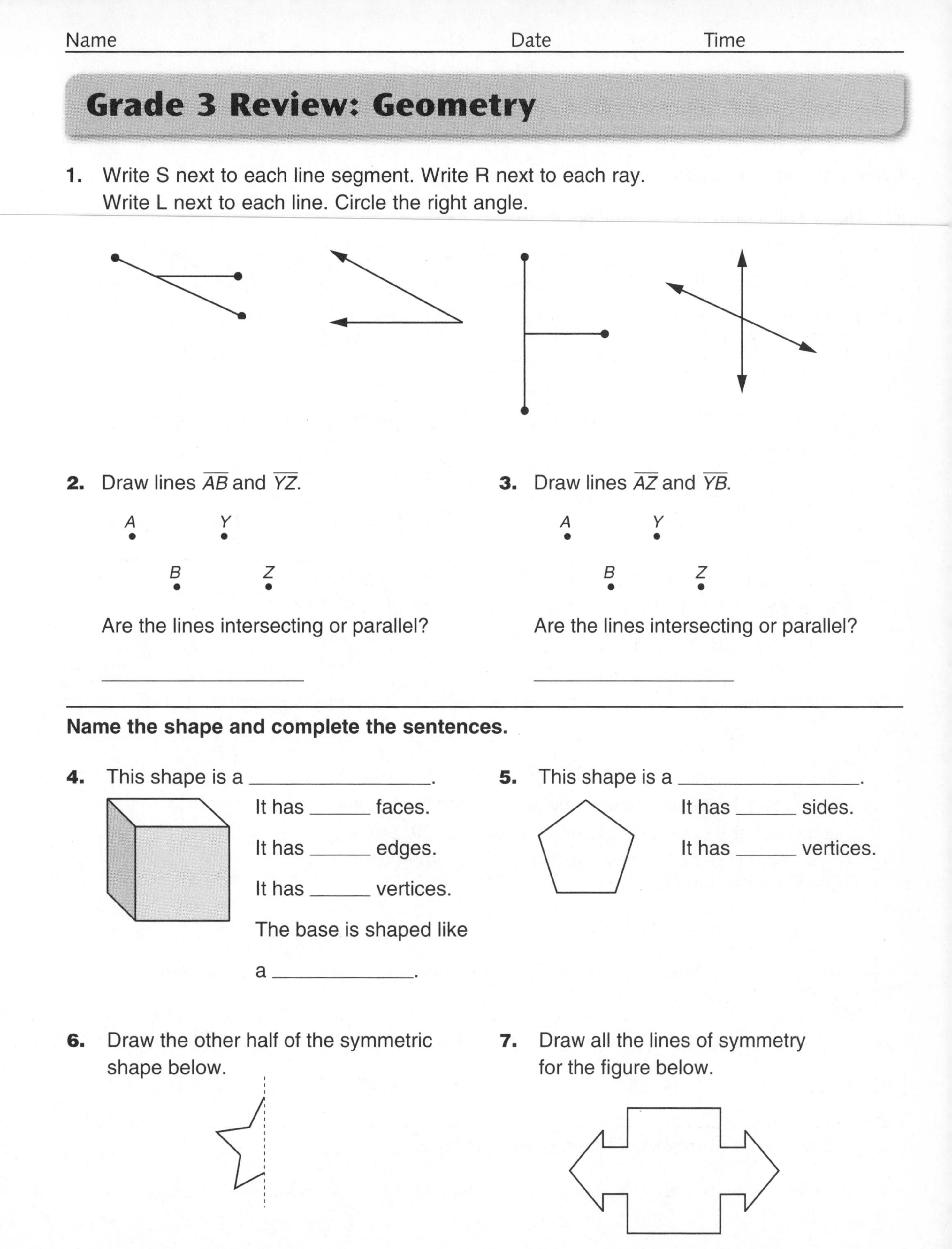

1. Write S next to each line segment. Write R next to each ray. Write L next to each line. Circle the right angle.

2. Draw lines $\overline{AB}$ and $\overline{YZ}$.

A Y

B Z

Are the lines intersecting or parallel?

____

3. Draw lines $\overline{AZ}$ and $\overline{YB}$.

A Y

B Z

Are the lines intersecting or parallel?

____

Name the shape and complete the sentences.

4. This shape is a ______________.

It has ______ faces.

It has ______ edges.

It has ______ vertices.

The base is shaped like a ______________.

5. This shape is a ______________.

It has ______ sides.

It has ______ vertices.

6. Draw the other half of the symmetric shape below.

7. Draw all the lines of symmetry for the figure below.

Name Date Time

Grade 3 Review: Patterns, Functions, and Algebra

Complete the frames-and-arrows diagrams.

1. Use a dollar sign and decimal point.

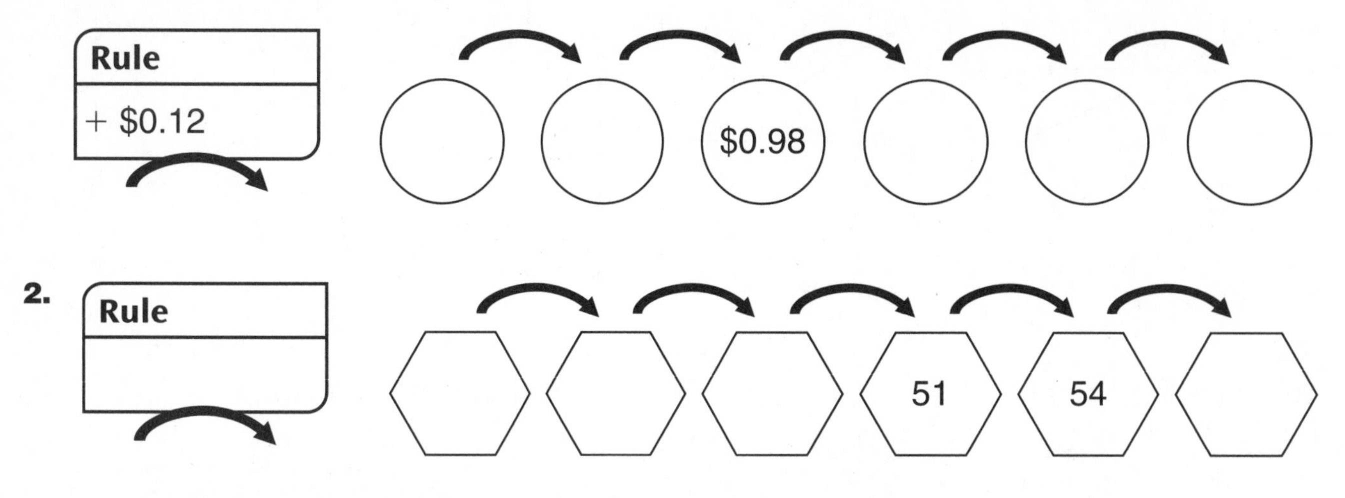

3. Describe the number pattern below. Write a number sentence that shows how to find the next number in the pattern.

0, 1, 1, 2, 3, 5, 8, 13

____

____

4. Bria earns $5.00 each week for helping around the house. Each week she buys a magazine that costs $3.00. She saves the rest of her money. Write a number sentence that shows how much money

Bria will have saved after 4 weeks. ____________________

Write <, >, or =.

5. 48 ÷ 8 ______ 6 × 8

6. (3 × 2) + 5 ______ 3 × (2 + 5)

7. 4 + (8 − 5) ______ (4 + 8) − 5

8. 15 − (5 × 3) ______ (15 − 5) × 3

Tell whether each number sentence is true or false.

9. (6 + 3) + 5 = (5 + 6) + 3 ____________

10. 6 + (9 + 7) = 16 + 7 ____________

Name Date Time

Practice Set 1

SRB 10–15

FACTS PRACTICE

For each problem set below, do as many problems as you can in one minute. You can ask someone to time you.

Problem Set 1

1. 8 + 7 = ________

2. 3 + 1 = ________

3. 9 − 3 = ________

4. 13 − 9 = ________

5. 8 − 4 = ________

6. 6 + 2 = ________

7. 12 − 8 = ________

8. 8 + 5 = ________

9. 10 − 2 = ________

10. 9 − 0 = ________

11. 6 + 6 = ________

12. 7 + 7 = ________

13. 5 + 8 = ________

14. 8 − 3 = ________

15. 18 − 9 = ________

Problem Set 2

16. 9 + 1 = ________

17. 11 − 4 = ________

18. 7 + 2 = ________

19. 5 + 6 = ________

20. 1 + 7 = ________

21. 9 − 5 = ________

22. 7 + 3 = ________

23. 8 − 4 = ________

24. 6 + 5 = ________

25. 9 + 1 = ________

26. 14 − 6 = ________

27. 4 + 8 = ________

28. 7 + 4 = ________

29. 14 − 2 = ________

30. 10 + 4 = ________

Problem Set 3

31. 9 − 6 = ________

32. 16 − 8 = ________

33. 4 + 5 = ________

34. 8 + 3 = ________

35. 2 + 5 = ________

36. 15 − 9 = ________

37. 9 − 7 = ________

38. 10 + 1 = ________

39. 7 − 7 = ________

40. 8 + 6 = ________

41. 5 + 5 = ________

42. 8 − 5 = ________

43. 2 + 9 = ________

44. 16 − 6 = ________

45. 14 − 9 = ________

Use with or after Lesson 1•1.

Name Date Time

Practice Set 2

SRB 1-4 91 149

Match each description with the correct example.
Write the letter that identifies the example.

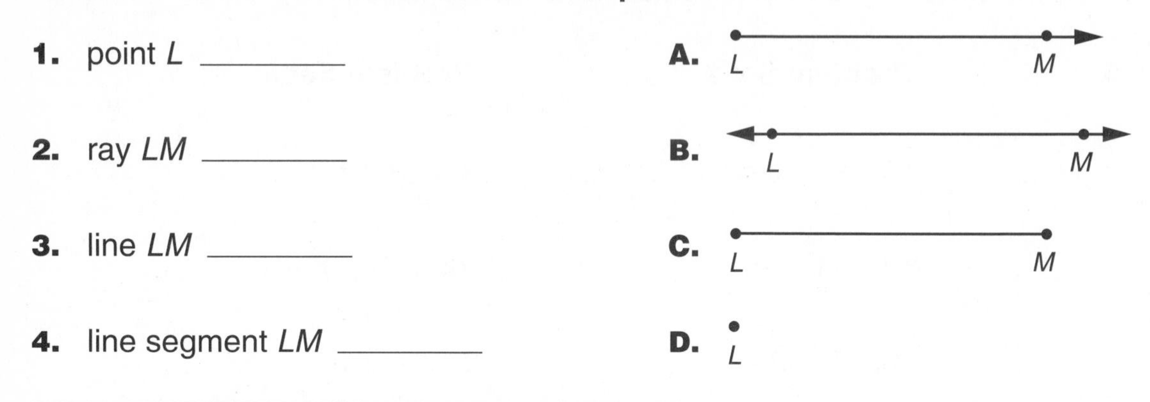

5. Use the clues to complete the puzzle.

1,000s	100s	10s	1s

- Write the result of 21 ÷ 7 in the ones place.
- Multiply 8 × 9. Subtract 65. Write the result in the tens place.
- Double the number in the ones place. Write the result in the thousands place.
- Divide 18 by 6. Add 5 and write the result in the hundreds place.

Fill in the name-collection boxes. Use as many different numbers and operations as you can.

Example

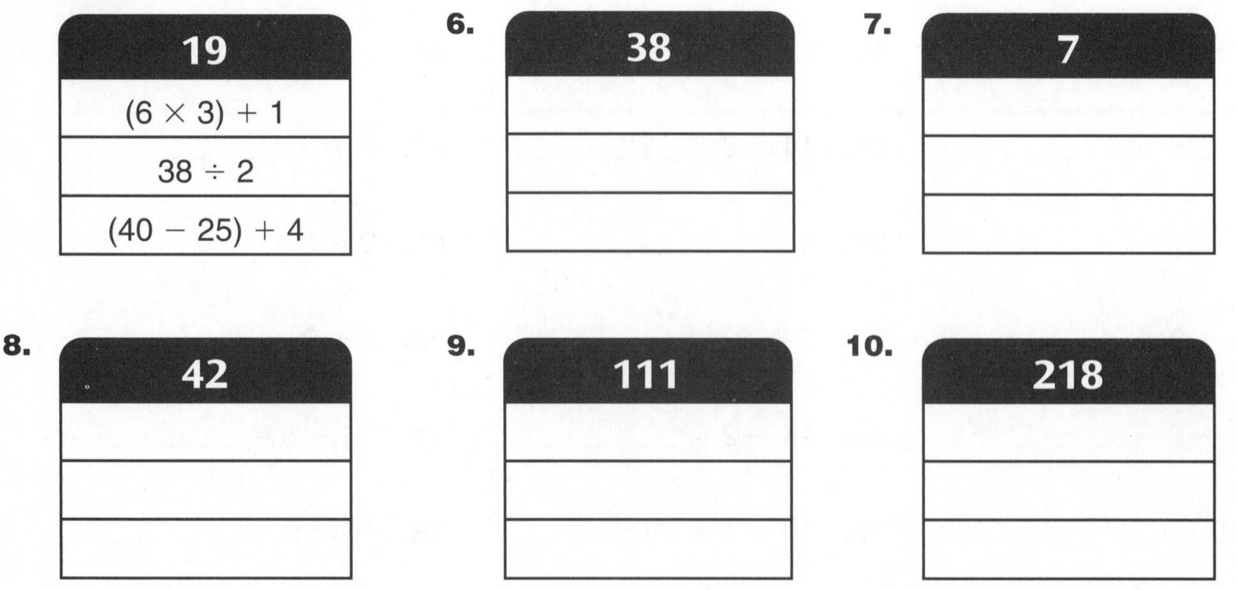

Name Date Time

Practice Set 2 *continued*

SRB 10–17 162–165

Complete the "What's My Rule?" tables and rule box.

11.

Rule
out = in × 3

in	out
3	9
4	12
7	
11	
15	

12.

Rule

in	out
8	4
14	7
	9
24	
36	18

Solve.

13. The Coffee-to-Go Cafe uses about 5 gallons of milk per day.

a. About how many gallons of milk does it use in a week (7 days)? ________ gallons

b. How many gallons in 5 weeks? ________ gallons

c. How many in one year (52 weeks)? ________ gallons

COMPUTATION PRACTICE **Solve.**

14. 18 − 17

15. 68 − 34

16. 74 + 27

17. 5,101 − 540

18. 500 − 290

19. 402 + 293

20. 3,418 + 6,583

21. 49 − 6

22. 120 − 30

23. 81 + 40

24. 35 − 22

25. 350 − 150

Name ______________________ Date __________ Time __________

Practice Set 3

SRB 6 87–99

Write your answers below.

1. Name the angle that has the smaller measure. __________

2. What is another name for $\angle GHI$? __________

3. What is the vertex of $\angle GHI$? __________

4. Which line segments make up $\angle ABC$? __________

5. What is the vertex of $\angle ABC$? __________

6. What is another name for $\angle ABC$? __________

7. What type of angle is $\angle EFG$? __________

8. Draw one more line segment to form a rectangle. Name the vertices. __________

9. **Writing/Reasoning** Compare $\angle ABC$ and $\angle GFH$. Which angle has the greater measure? How do you know?

____

____

Write <, >, or =.

10. 6,985 ______ 6,367

11. 459 ______ 489

12. 1,640 ______ 1,643

13. 10,387 ______ 10,340

Name Date Time

Practice Set 4

SRB 99 100 150 151

Match each name with the correct figure.
Write the letter that identifies the figure.

A. **B.** **C.** **D.**

1. rhombus ________

2. trapezoid ________

3. square ________

4. kite ________

5. **Writing/Reasoning** Which of these shapes has two names? Explain your answer.

____

____

____

COMPUTATION PRACTICE

Solve.

6. 2×4

7. $14 - 7$

8. $84 - 27$

9. $300 + 500$

10. $35 - 19$

11. $30 + 83$

12. $43 + 21$

13. 9×0

14. $(50 + 20) \times 4 =$ ________

15. $27 - (5 \times 4) =$ ________

16. $200 + 150 + 100 =$ ________

17. $500 + 440 + 120 =$ ________

Name Date Time

Practice Set 4 *continued*

SRB 160 161

Complete the frames-and-arrows problems.

Example

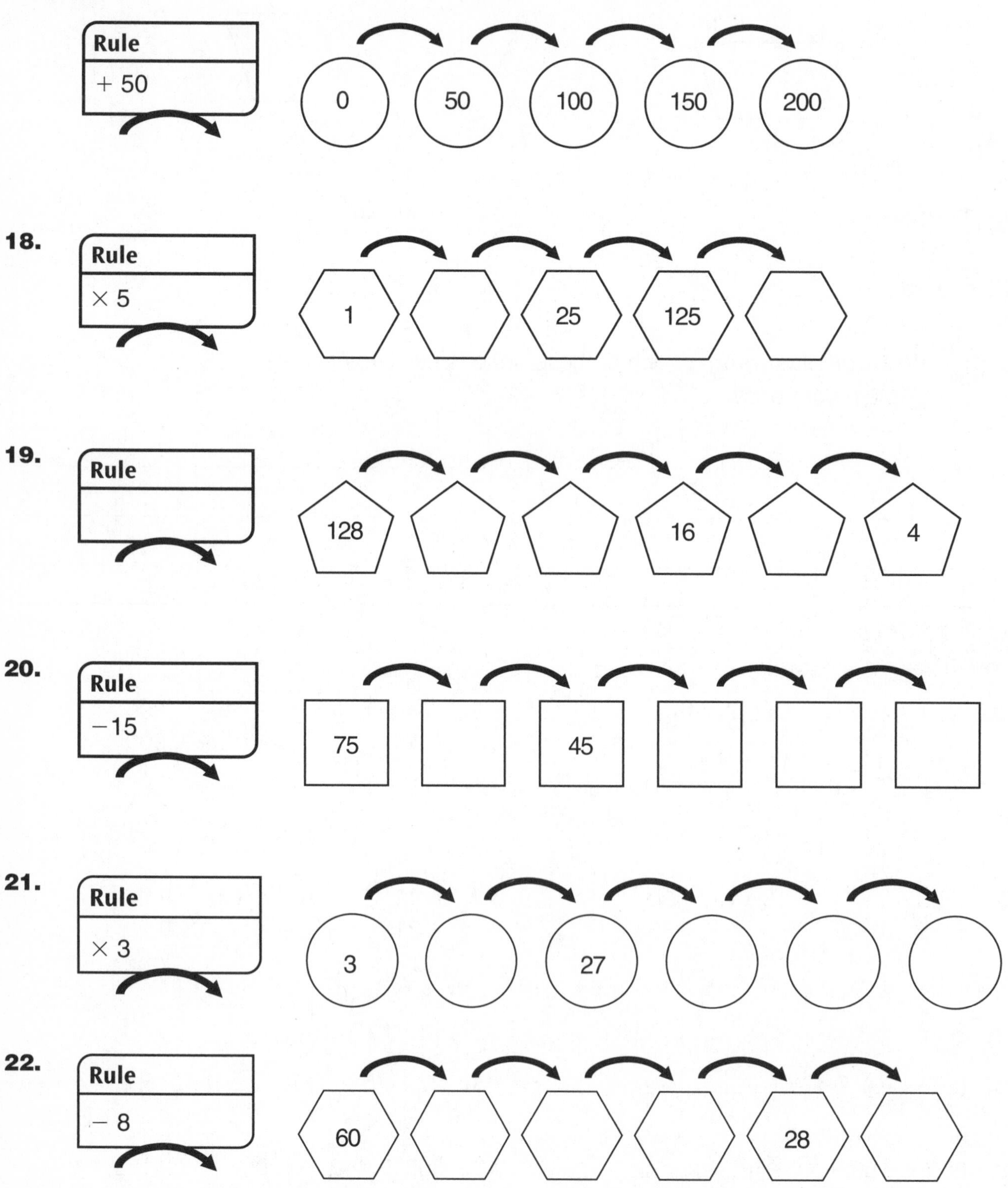

Use with or after Lesson 1•4.

Name Date Time

Practice Set 5

SRB 96 97 150 151

1. Which figures are polygons? ______

A. B. C.

D. E. F.

Write 2 addition and 2 subtraction facts for each group of numbers.

2. 2, 9, 11

3. 8, 9, 17

4. 1, 6, 7

5. 3, 4, 7

6. 4, 6, 10

7. 6, 7, 13

Write the number models with parentheses and then solve.

8. Add 15 to the difference of 105 and 70. ______

9. Subtract the sum of 8 and 3 from 18. ______

10. Add 9 to the difference of 50 and 16. ______

11. Subtract the sum of 81 and 42 from 338. ______

Name Date Time

Practice Set 5 *continued*

SRB 16–24

FACTS PRACTICE

For each problem set below, do as many problems as you can in one minute. You can ask someone to time you.

Problem Set 1

12. $18 \div 6 =$ ______

13. $7 \times 7 =$ ______

14. $4 \times 5 =$ ______

15. $54 \div 9 =$ ______

16. $3 \times 4 =$ ______

17. $64 \div 8 =$ ______

18. $6 \times 8 =$ ______

19. $6 \times 6 =$ ______

20. $9 \times 9 =$ ______

21. $49 \div 7 =$ ______

22. $9 \times 4 =$ ______

23. $7 \times 9 =$ ______

24. $6 \times 2 =$ ______

25. $56 \div 7 =$ ______

26. $7 \times 6 =$ ______

Problem Set 2

27. $4 \times 8 =$ ______

28. $36 \div 6 =$ ______

29. $5 \times 9 =$ ______

30. $8 \times 4 =$ ______

31. $3 \times 8 =$ ______

32. $48 \div 6 =$ ______

33. $7 \times 4 =$ ______

34. $28 \div 7 =$ ______

35. $2 \times 5 =$ ______

36. $6 \times 3 =$ ______

37. $32 \div 4 =$ ______

38. $6 \times 7 =$ ______

39. $4 \times 7 =$ ______

40. $20 \div 5 =$ ______

41. $8 \times 9 =$ ______

Problem Set 3

42. $9 \times 6 =$ ______

43. $24 \div 4 =$ ______

44. $2 \times 7 =$ ______

45. $8 \times 7 =$ ______

46. $3 \times 7 =$ ______

47. $27 \div 3 =$ ______

48. $7 \times 5 =$ ______

49. $3 \times 6 =$ ______

50. $8 \div 4 =$ ______

51. $4 \times 3 =$ ______

52. $9 \times 8 =$ ______

53. $63 \div 9 =$ ______

54. $9 \times 5 =$ ______

55. $15 \div 3 =$ ______

56. $2 \times 8 =$ ______

Name Date Time

Practice Set 6

SRB 10–17 96 97

1. Which figures are regular polygons? ______________

A. **B.** **C.**

D. **E.** **F.**

2. **Writing/Reasoning** Explain why the figures you chose in Problem 1 are regular polygons.

____

____

COMPUTATION PRACTICE

Solve.

3. 120 − ________ = 30

4. 82 + 17 = ________

5. 70 + 26 = ________

6. 40 − 17 = ________

7. 16 − 8 = ________

8. 9 × 9 = ________

9. $\begin{array}{r} 87 \\ -\ 36 \\ \hline \end{array}$

10. $\begin{array}{r} 33 \\ -\ 14 \\ \hline \end{array}$

11. $\begin{array}{r} 37 \\ \times\ 8 \\ \hline \end{array}$

12. $\begin{array}{r} 11 \\ +\ 34 \\ \hline \end{array}$

13. $\begin{array}{r} 120 \\ \times\ 50 \\ \hline \end{array}$

14. $\begin{array}{r} 521 \\ +\ 131 \\ \hline \end{array}$

15. $\begin{array}{r} 34 \\ +\ 15 \\ \hline \end{array}$

16. $\begin{array}{r} 35 \\ \times\ 3 \\ \hline \end{array}$

Name Date Time

Practice Set 7

SRB 20 160 161

Match each term with the correct figure.
Write the letter that identifies the figure.

1. concentric circles ________ **A.**

2. right triangle ________ **B.**

3. rhombus ________ **C.**

4. rectangle ________ **D.**

Solve.

5. Joy wants to have enough balloons for her 22 party guests.

How many packages of 6 does she need? ____________________

6. Larry has a large pizza to share with 3 friends. If the pizza is divided into 16 slices, how many slices will each person, including Larry, get?

____

7. If four more friends join Larry and the others, how many slices will each person get?____________________

Complete the frames-and-arrows problem.

8.

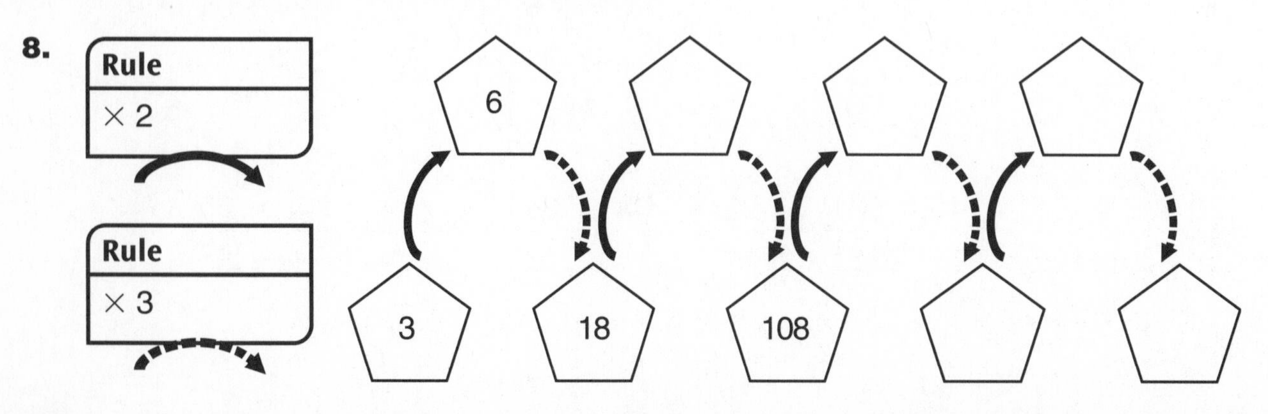

Name ______ Date ______ Time ______

Practice Set 8

SRB 30 31 133 134

Find the area, in square units, of each rectangle. Then write the number model.

Reminder: Area = length (l) × width (w)

Example

$6 \times 7 = 42$

1. ______

2. ______

Write the digit in the hundredths place for each number.

3. 5.925 ______ **4.** 1.043 ______ **5.** 8.100 ______ **6.** 0.280 ______ **7.** 3.313 ______

Write two multiplication and two division facts for each group of numbers.

8. 6, 7, 42

______
______
______
______

9. 3, 9, 27

______
______
______
______

10. 4, 8, 32

______
______
______
______

11. 5, 9, 45

______
______
______
______

12. 2, 8, 16

______
______
______
______

13. 4, 7, 28

______
______
______
______

Write a number model.

14. Erin has 3 sets of 25 seashells. How many seashells does Erin have in all?

______

Name Date Time

Practice Set 9

SRB 144

Use the map of Lakeville to answer the questions.

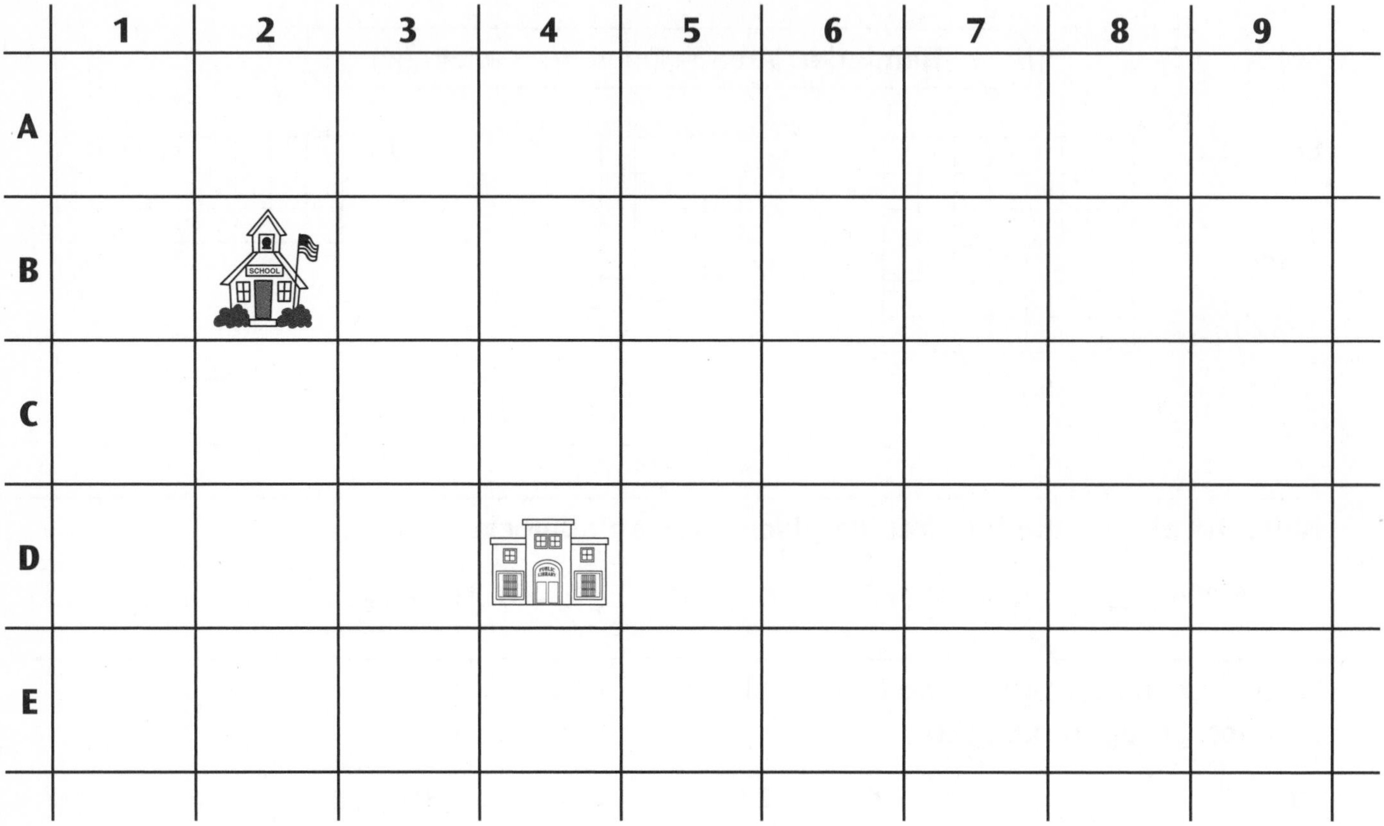

	1	2	3	4	5	6	7	8	9
A									
B		SCHOOL							
C									
D				PUBLIC LIBRARY					
E									

1. In what region can you find the school? _________

2. In what region can you find the library? _________

3. Amy lives in region E-6. Draw her house on the map.

4. **Writing/Reasoning** Does Amy live closer to the school or to the library? Explain your answer.

____

____

5. Draw and label a store on the map. In what region did you draw the store? _________

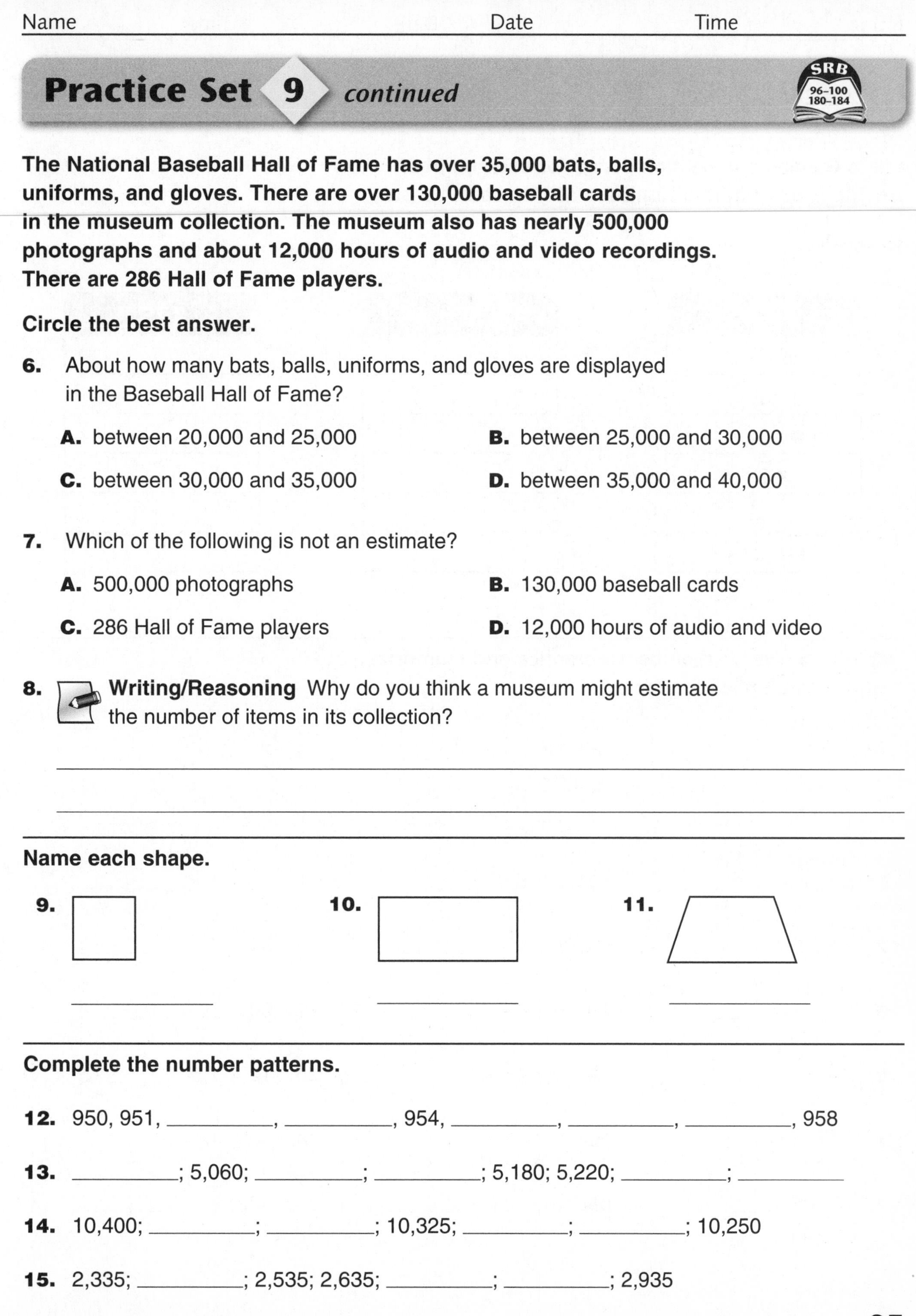

Name Date Time

Practice Set 9 *continued*

SRB 96–100 180–184

The National Baseball Hall of Fame has over 35,000 bats, balls, uniforms, and gloves. There are over 130,000 baseball cards in the museum collection. The museum also has nearly 500,000 photographs and about 12,000 hours of audio and video recordings. There are 286 Hall of Fame players.

Circle the best answer.

6. About how many bats, balls, uniforms, and gloves are displayed in the Baseball Hall of Fame?

A. between 20,000 and 25,000 **B.** between 25,000 and 30,000

C. between 30,000 and 35,000 **D.** between 35,000 and 40,000

7. Which of the following is not an estimate?

A. 500,000 photographs **B.** 130,000 baseball cards

C. 286 Hall of Fame players **D.** 12,000 hours of audio and video

8. **Writing/Reasoning** Why do you think a museum might estimate the number of items in its collection?

____

____

Name each shape.

9. ______________ **10.** ______________ **11.** ______________

Complete the number patterns.

12. 950, 951, ________, ________, 954, ________, ________, ________, 958

13. ________; 5,060; ________; ________; 5,180; 5,220; ________; ________

14. 10,400; ________; ________; 10,325; ________; ________; 10,250

15. 2,335; ________; 2,535; 2,635; ________; ________; 2,935

Name Date Time

Practice Set 10

SRB 16–24 149

Fill in the name-collection boxes for each number.
Use as many different numbers as you can.

Example

42
50 − 8
20 × 2 + 2
100 ÷ 4 + 17
3 × 2 × 7
84/2
70 ÷ 10 × 6
1.
27
2.
125

FACTS PRACTICE

Solve. Remember to practice and memorize your multiplication facts.

3. $6 \times 6 =$ ________ **4.** $12 \times 6 =$ ________ **5.** $7 \times 5 =$ ________

6. $10 \times 7 =$ ________ **7.** $11 \times 4 =$ ________ **8.** $9 \times 9 =$ ________

9. $6 \times 8 =$ ________ **10.** $7 \times 5 =$ ________ **11.** $9 \times 11 =$ ________

12. $2 \times 12 =$ ________ **13.** $11 \times 8 =$ ________ **14.** $8 \times 12 =$ ________

15. $10 \times 11 =$ ________ **16.** $8 \times 4 =$ ________ **17.** $60 \div 10 =$ ________

18. $9 \times 5 =$ ________ **19.** $7 \times 11 =$ ________ **20.** $12 \times 7 =$ ________

21. $16 \div 8 =$ ________ **22.** $15 \div 3 =$ ________ **23.** $60 \div 12 =$ ________

24. $12 \div 3 =$ ________ **25.** $10 \div 2 =$ ________ **26.** $40 \div 8 =$ ________

Name Date Time

Practice Set 11

SRB 4 10 29–31

COMPUTATION PRACTICE

Add.

1.
$$\begin{array}{r} 700 \\ 40 \\ +\quad 2 \\ \hline \end{array}$$

2.
$$\begin{array}{r} 8{,}000 \\ 200 \\ 30 \\ +\quad 1 \\ \hline \end{array}$$

3.
$$\begin{array}{r} 60{,}000 \\ 500 \\ 70 \\ +\quad 3 \\ \hline \end{array}$$

4.
$$\begin{array}{r} 90{,}000 \\ 6{,}000 \\ 800 \\ 10 \\ +\quad 4 \\ \hline \end{array}$$

5.
$$\begin{array}{r} 800{,}000 \\ 3{,}000 \\ 300 \\ +\quad 1 \\ \hline \end{array}$$

6.
$$\begin{array}{r} 4{,}000{,}000 \\ 900{,}000 \\ 30{,}000 \\ 400 \\ +\quad 90 \\ \hline \end{array}$$

7. Use the clues to complete the place-value puzzle.

- Divide 72 by 6. Subtract 4 and write the result in the ones place.
- Double the number in the ones place and divide by 8. Write the result in the tens place.
- Multiply 9 × 10. Subtract 83. Write the result in the hundreds place.
- Halve the number in the tens place. Multiply by 3 and write the result in the thousands place.
- Divide 27 by the number in the thousands place. Write the result in the ten-thousands place.

10,000s	1,000s	100s	10s	1s

Name _______________ Date _______________ Time _______________

Practice Set 11 *continued*

SRB 16-21

FACTS PRACTICE

For each problem set below, do as many problems as you can in one minute. You can ask someone to time you.

Problem Set 1

8. $10 \times 6 =$ ________

9. $7 \times 11 =$ ________

10. $48 \div 4 =$ ________

11. $2 \times 9 =$ ________

12. $10 \times 4 =$ ________

13. $90 \div 9 =$ ________

14. $5 \times 11 =$ ________

15. $10 \times 12 =$ ________

16. $49 \div 7 =$ ________

17. $4 \times 12 =$ ________

18. $7 \times 8 =$ ________

19. $63 \div 9 =$ ________

20. $45 \div 5 =$ ________

21. $6 \times 7 =$ ________

22. $18 \div 2 =$ ________

Problem Set 2

23. $2 \times 5 =$ ________

24. $9 \times 11 =$ ________

25. $110 \div 10 =$ ________

26. $8 \times 4 =$ ________

27. $11 \times 10 =$ ________

28. $6 \times 12 =$ ________

29. $11 \times 11 =$ ________

30. $12 \times 11 =$ ________

31. $4 \times 7 =$ ________

32. $12 \times 16 =$ ________

33. $63 \div 7 =$ ________

34. $8 \times 6 =$ ________

35. $3 \times 3 =$ ________

36. $45 \div 9 =$ ________

37. $16 \div 4 =$ ________

Problem Set 3

38. $54 \div 9 =$ ________

39. $12 \times 2 =$ ________

40. $12 \times 5 =$ ________

41. $12 \times 3 =$ ________

42. $121 \div 11 =$ ________

43. $60 \times 5 =$ ________

44. $8 \times 7 =$ ________

45. $144 \div 12 =$ ________

46. $8 \div 4 =$ ________

47. $33 \div 3 =$ ________

48. $42 \div 7 =$ ________

49. $81 \div 9 =$ ________

50. $32 \div 8 =$ ________

51. $18 \div 9 =$ ________

52. $7 \times 3 =$ ________

Name _______________ Date _______________ Time _______________

Practice Set 12

SRB 4 10–20 34–37

Use digits to write the following numbers.

1. twenty-four thousand, nine hundred sixty-eight _______________

2. seventy-six thousand, six hundred fourteen _______________

3. six thousand, nine hundred two _______________

Write the number words for the following numbers.

4. 12,743 _______________

5. 8,054 _______________

6. 69,231 _______________

7. 4,782 _______________

COMPUTATION PRACTICE

Solve.

8. $26 + 47$

9. $63 + 18$

10. 16×4

11. 180×7

12. 196×0

13. $32.1 + 18.7$

14. $1.25 + 6.43$

15. $8.40 - 5.01$

16. 11×9

17. $85 - 38$

18. $20 + 83 + 17$

19. $83 - 41$

Name _______________ Date _______________ Time _______________

Practice Set 13

SRB 4 26–31 71–73

The tally chart at the right shows the number of items that some fourth graders missed on a quiz.

Number of Items Missed	Number of Students
0	HHT //
1	HHT /
2	///
3	//
4	//
5	/
6	//
7	/

1. How many students reported the number of items they missed? _______________

2. What is the maximum (greatest) number of items missed? _______________

3. What is the minimum (least) number of items missed? _______________

4. What is the range? _______________

5. What is the mode (most frequent) number of items missed? _______________

Use digits to write the following numbers.

6. sixteen thousand, five hundred forty-seven

____

7. eight and two tenths

____

8. seven and nine tenths

____

Write the number words for the following numbers.

9. 21,894 ____

10. 14.1 ____

11. 48,563 ____

12. 903 ____

Name Date Time

Practice Set 13 *continued*

SRB 150 151 162–164

Fill in the "What's My Rule?" tables and rule boxes.

13.

Rule
out = in * 20

in	out
9	180
12	240
15	
25	
100	

14.

Rule

in	out
7	3.5
10	6.5
	10.5
16.5	
20.5	17

15.

Rule

in	out
80	20
160	
	90
2,400	
4,800	1,200

16.

Rule
out = in * 10

in	out
3	
6	
	90
	120
15	

Rewrite the number sentences with parentheses to make them correct.

17. 6 * 11 − 7 = 59

____

18. 2.2 = 8 − 3 + 2.8

____

19. 330 − 150 − 60 = 240

____

20. 18 = 2 * 5.4 + 3.6

____

21. 7 * 2.1 + 5 * 12 = 74.7

____

22. 230 = 4 * 60 − 10

____

23. 3 * 9 + 3 − 4 = 32

____

24. 584 = 11 * 50 + 34

____

25. 12 ÷ 4 + 2 * 3 = 6

____

26. 6 + 3 * 2 − 4 = 14

____

Practice Set 14

Mr. Adema asked his piano students to estimate the number of hours they practice each week. The tally chart shows the data he collected. Use the table to help you answer the questions below.

Number of Hours	Number of Students
2	//
3	~~////~~ //
4	////
5	///
6	//
7	/
8	/

1. Construct a line plot for the data.

2. What is the maximum number of hours spent practicing each week? ____________

3. What is the minimum number of hours spent practicing each week? ____________

4. What is the range? ____________

5. What is the median number of hours? ____________

6. **Writing/Reasoning** What is the mean number of hours spent practicing? Explain how you found your answer.

____

____

Complete the frames-and-arrows problems. Write the missing rule.

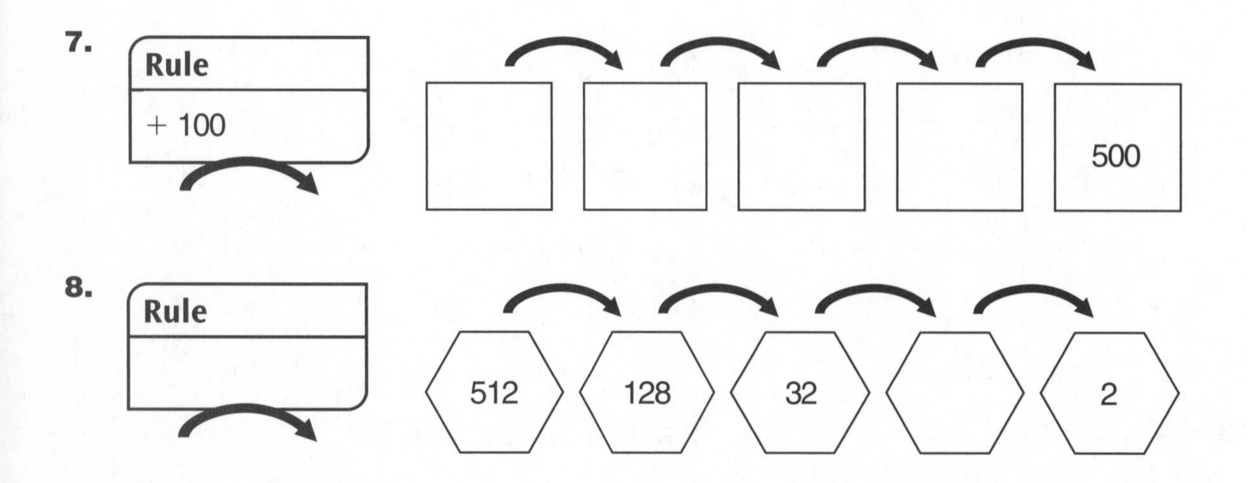

Name Date Time

Practice Set 14 *continued*

SRB 4 29-31 149 153

9. Use the clues to complete the place-value puzzle.

- Divide 88 by 11. Add 1 and write the result in the thousands place.
- Double the number in the thousands place and divide by 3. Write the result in the tens place.
- Multiply 4 * 12. Subtract 42. Write the result in the hundreds place.
- Divide 63 by the number in the thousands place. Write the result in the ones place.
- Halve the number in the tens place. Add 1 and write the result in the ten-thousands place.

10,000s	1,000s	100s	10s	1s

Fill in the name-collection boxes. Use as many different numbers and operations as you can.

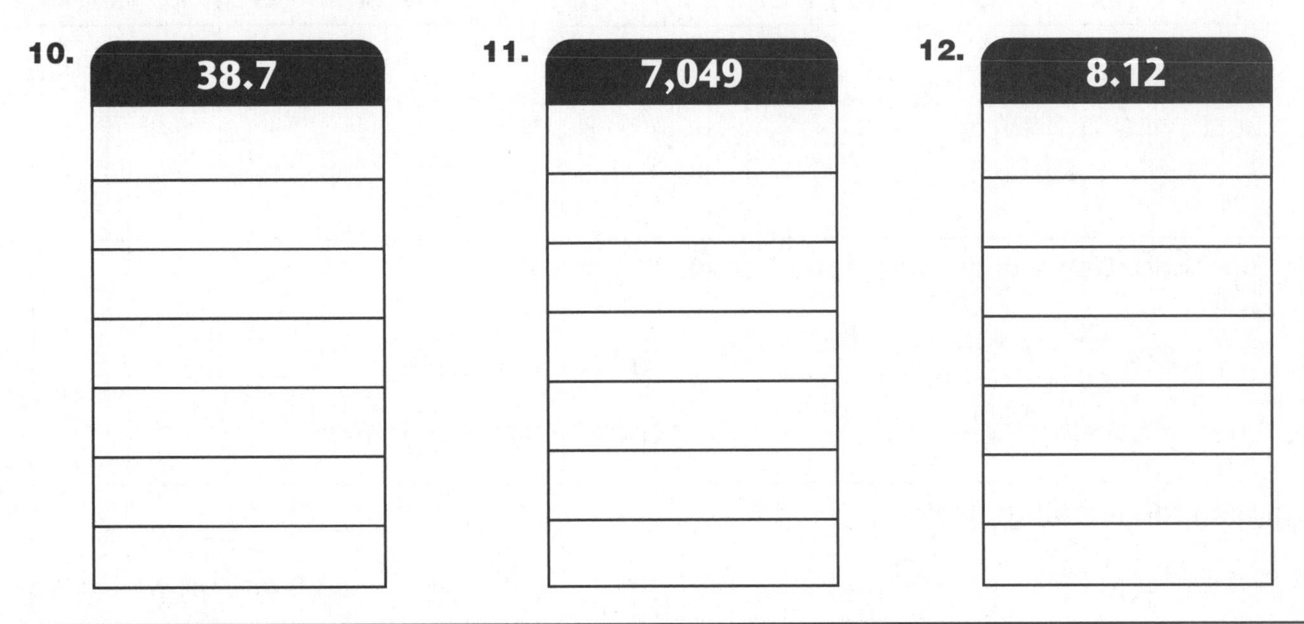

Write as a multiplication problem. Then find the product.

13. 2 + 2 + 2

________ * ________ = ________

14. 8 + 8 + 8 + 8 + 8

________ * ________ = ________

15. 10 + 10

________ * ________ = ________

16. 9 + 9 + 9 + 9 + 9 + 9 + 9

________ * ________ = ________

Name Date Time

Practice Set 15

SRB 10 11 180–183

COMPUTATION PRACTICE

Add. Use the partial-sums method.

1.

100s	10s	1s
2	0	4
+ 1	4	9

2.

100s	10s	1s
5	5	1
+ 2	6	7

3.

1000s	100s	10s	1s
	8	5	9
+ 1	5	9	6

Add. Use the column addition method.

4.

100s	10s	1s
7	3	4
+ 4	7	8

5.

100s	10s	1s
5	9	2
+ 8	7	9

6.

1000s	100s	10s	1s
2	7	3	5
+ 1	3	0	5

COMPUTATION PRACTICE

Add. Use any method you choose.

7. 795 + 616 = ______________

8. 8,214 + 5,488 = ______________

9. 5,838 + 8,956 = ______________

10. 50,694 + 39,518 = ______________

Complete the missing factors.

11. 7 * _____ = 21

12. _____ * 4 = 36

13. _____ * 8 = 64

14. 12 * _____ = 96

15. 400 * _____ = 3,600

16. _____ * 5 = 350

Write a number model to show how to estimate the total cost.

17. 12 rulers that cost $1.05 each ______________________

18. 4 scissors that cost $0.69 each ______________________

19. 7 books that cost $3.45 each ______________________

Use with or after Lesson 2•7.

Name Date Time

Practice Set 16

SRB 73 76 128–130

Mrs. Lewis made a graph to show the number of art projects the students have completed. Use the bar graph to find the following landmarks for the data.

Art Projects Completed by Students

Number of Completed Projects	Number of Students
1	4
2	1
3	5
4	3
5	2

1. What is the maximum number of completed projects? ____________
2. What is the minimum number of completed projects? ____________
3. What is the range? ____________
4. What is the median? ____________

Measure the line segments to the nearest centimeter.

5. ______________________________

   ____________ cm

6. ________________________

   ____________ cm

7. ___________________________

   ____________ cm

8. _____________

   ____________ cm

Name Date Time

Practice Set 16 *continued*

SRB 160 161

Complete the frames-and-arrows problems.

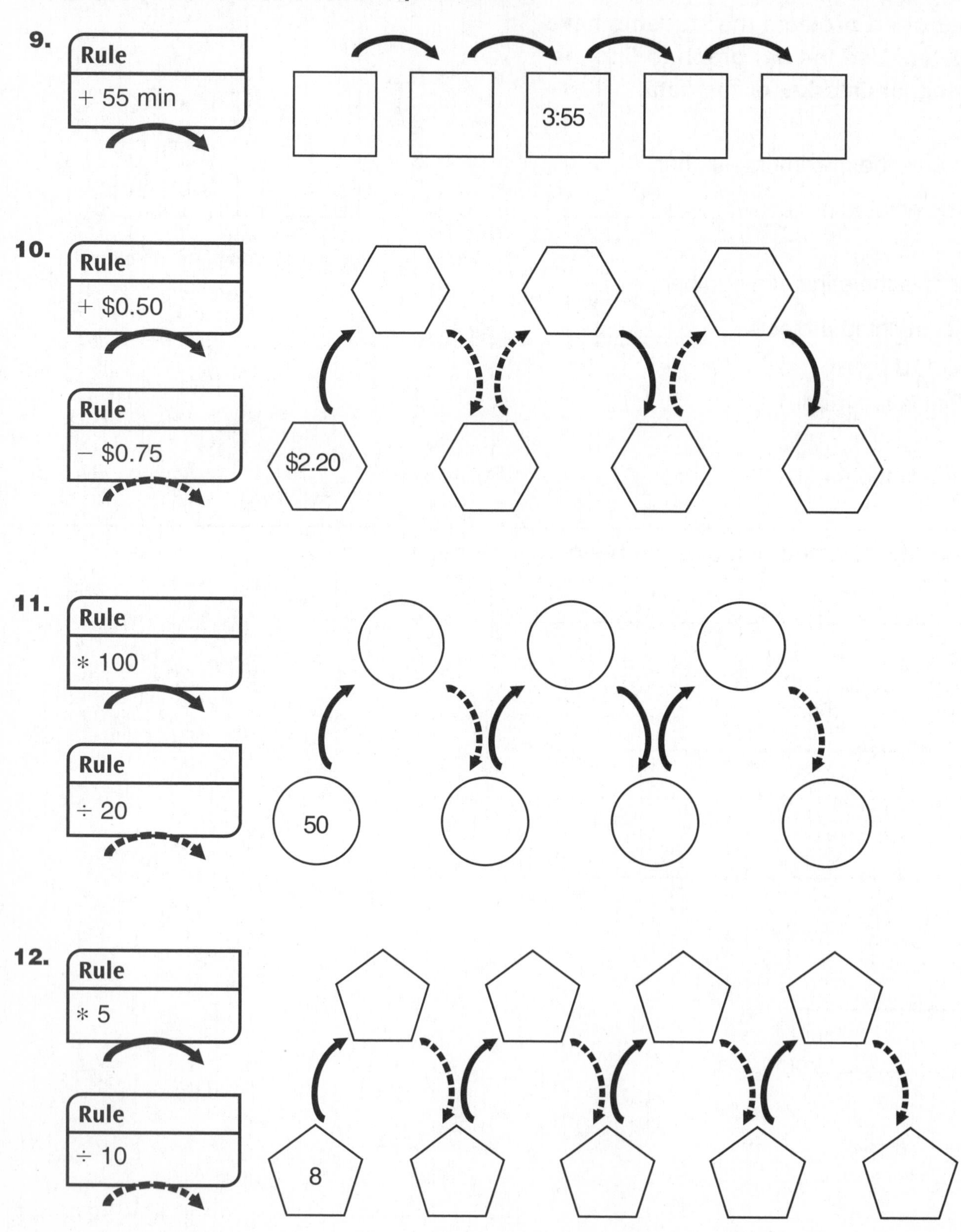

Name Date Time

Practice Set 17

SRB 9 12–15 162–164

COMPUTATION PRACTICE

Subtract. Use the trade-first method.

1.

100s	10s	1s
	9	2
−	3	7

2.

100s	10s	1s
6	2	4
− 2	8	6

3.

100s	10s	1s
3	4	8
− 1	5	9

Subtract. Use the partial-differences method.

4.

100s	10s	1s
	7	8
−	5	7

5.

100s	10s	1s
6	5	8
− 2	7	0

6.

100s	10s	1s
7	3	4
− 3	8	6

Subtract. Use any method you choose.

7. 79 − 21 = _____ **8.** 33 − 17 = _____ **9.** 636 − 498 = _____

Solve mentally. Use the counting-up strategy.

10. 961 − 185 = _____ **11.** 70 − 51 = _____ **12.** 130 − 97 = _____

Write the missing numbers.

13. 9, 17, _____, 33, _____, _____, _____, 65

14. 0.2, _____, _____, 0.8, _____, _____, _____, _____

15. $\frac{1}{7}$, $\frac{2}{7}$, _____, $\frac{4}{7}$, _____, _____, 1, _____, _____

Name Date Time

Practice Set 17 *continued*

SRB 10–24 41–44

Write your answers below.

16. How many pieces of fruit are there? ____________

17. What fraction of the fruit is apples? ____________

18. What fraction of the fruit is pears? ____________

19. What fraction of the fruit is bananas? ____________

20. **Writing/Reasoning** Explain how you found your answer to Problem 19.

____

____

COMPUTATION PRACTICE **Solve.**

21. 67 * 4

22. 53 * 8

23. 84 ÷ 3 = ________

24. 675 * 6

25. 8,229 + 3,160

26. 7,583 − 5,432

27. 3,499 * 0

28. 673 − 64

29. 467 + 185

30. 320 * 8

31. 560 ÷ 8 = ________

32. 8,524 − 1,996

Use with or after Lesson 2•9.

Name Date Time

Practice Set 18

SRB 80–86 162–166

Complete the "What's My Rule?" tables and rule boxes.

1. in → Rule: [blank] → out

in	out
9	3
18	6
27	
36	

2. in → Rule: Multiply by 5 → out

in	out
8	
10	
12	
14	

3. in → Rule: [blank] → out

in	out
10	40
20	80
30	
	160

4. in → Rule: Multiply by 8 → out

in	out
	32
	48
	64
	80

Jamie and Allison each have a bag of 12 colored cubes. Some cubes are blue, some cubes are red, and some cubes are green. For Problems 5 and 6, use the probability clues to tell how many cubes of each color are in each bag.

5. From Jamie's bag, you have an equal chance of picking a red cube, a green cube, or a blue cube.

_______ blue cubes _______ red cubes _______ green cubes

6. From Allison's bag, you have an equal chance of picking a red cube or a blue cube, but there is only a 1 out of 6 chance of picking a green cube.

_______ blue cubes _______ red cubes _______ green cubes

7. **Writing/Reasoning** Explain how you found your answer to Problem 6.

____

____

Name Date Time

Practice Set 19

SRB 16–24

FACTS PRACTICE

For each problem set below, do as many problems as you can in one minute. You can ask someone to time you.

Problem Set 1

1. 9 * 6 = ________
2. 7 * 7 = ________
3. 21 ÷ 7 = ________
4. 12 * 8 = ________
5. 44 ÷ 4 = ________
6. 2 * 10 = ________
7. 11 * 4 = ________
8. 64 ÷ 8 = ________
9. 12 * 5 = ________
10. 10 * 11 = ________
11. 81 ÷ 9 = ________
12. 54 ÷ 6 = ________
13. 9 * 7 = ________
14. 48 ÷ 8 = ________
15. 9 * 3 = ________

Problem Set 2

16. 12 * 7 = ________
17. 2 * 9 = ________
18. 6 * 6 = ________
19. 9 * 9 = ________
20. 121 ÷ 11 = ________
21. 6 * 7 = ________
22. 4 * 12 = ________
23. 21 ÷ 3 = ________
24. 108 ÷ 9 = ________
25. 8 * 4 = ________
26. 42 ÷ 6 = ________
27. 144 ÷ 12 = ________
28. 4 * 10 = ________
29. 11 * 11 = ________
30. 8 * 3 = ________

Problem Set 3

31. 12 * 11 = ________
32. 54 ÷ 9 = ________
33. 42 ÷ 7 = ________
34. 7 * 3 = ________
35. 12 * 4 = ________
36. 55 ÷ 5 = ________
37. 6 * 8 = ________
38. 3 * 11 = ________
39. 90 ÷ 9 = ________
40. 48 ÷ 6 = ________
41. 12 * 9 = ________
42. 4 * 7 = ________
43. 3 * 8 = ________
44. 132 ÷ 12 = ________
45. 49 ÷ 7 = ________

Use with or after Lesson 3•2.

Name Date Time

Practice Set 19 *continued*

SRB 152 153 175

Write your answers below.

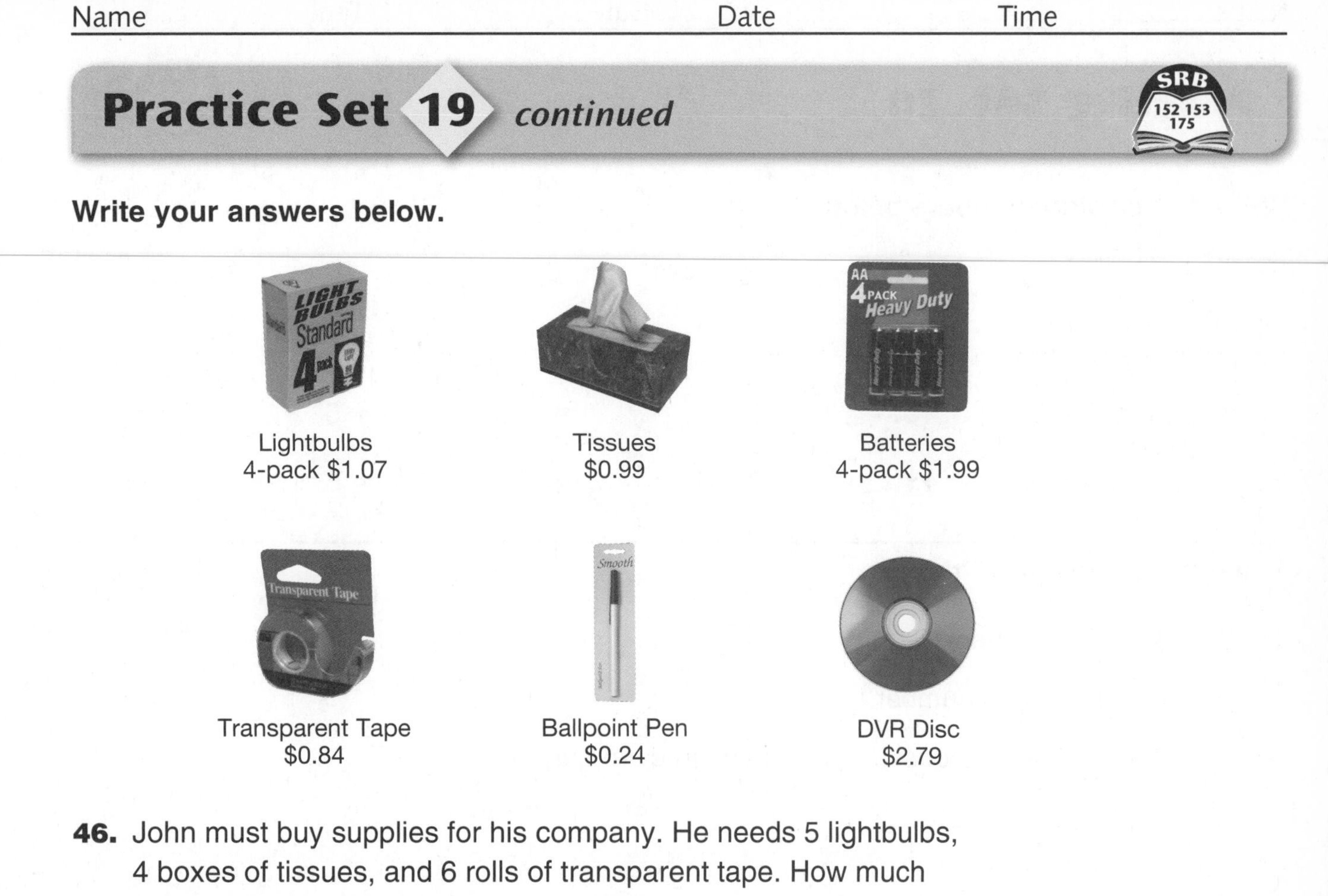

Lightbulbs 4-pack $1.07

Tissues $0.99

Batteries 4-pack $1.99

Transparent Tape $0.84

Ballpoint Pen $0.24

DVR Disc $2.79

46. John must buy supplies for his company. He needs 5 lightbulbs, 4 boxes of tissues, and 6 rolls of transparent tape. How much money does he need for these supplies?

____

47. Ms. Larson has $4.00. How many pens can she buy?

____

48. Judy and Sarah are going to record their school's play. They need 8 batteries and 2 DVR discs for the camera. They have $10.00. Do they have enough money to buy what they need? What is the difference between the money they have and the money they need?

____

49. About how much is each of the lightbulbs in the 4-pack?

____

50. About how much is each of the batteries in the 4-pack?

____

Name ______ Date ______ Time ______

Practice Set 20

SRB 46 16–24

Write the missing numbers below.

1. 3 * ______ = 15 **2.** ______ * 8 = 48 **3.** ______ / 7 = 6

4. 3 * ______ = 30 **5.** 18 / ______ = 9 **6.** ______ * 7 = 63

7. ______ * 4 = 32 **8.** ______ / 2 = 7 **9.** 20 / ______ = 5

10. ______ * 5 = 25 **11.** ______ / 7 = 4 **12.** 54 / ______ = 6

Solve the "Who am I?" riddles.

13. Clue 1: I am less than 10.

Clue 2: I am an odd number.

Clue 3: If you turn me upside down, I am an even number.

______

14. Clue 1: I am less than 100.

Clue 2: The sum of my digits is 8.

Clue 3: If you divide me by 2, I am an even number.

Clue 4: My tens digit and my ones digit are the same.

______

15. Clue 1: I am a number between 75 and 150.

Clue 2: My tens digit is three times my ones digit.

Clue 3: The sum of my digits is 5.

Clue 4: My hundreds digit and my ones digit are the same.

______

16. **Writing/Reasoning** Make up your own "Who am I?" riddle. Include the answer to your riddle.

______

______

Name Date Time

Practice Set 21

SRB 7 15 19 71 72 74

Find the missing number for each Fact Triangle. Then write the fact family for that triangle.

1. Fact Triangle: top 63; bottom: ___?___, 7; *, /

Missing number: ____

Fact family:

____

____

____

____

2. Fact Triangle: top ___?___; bottom: 8, 6; *, /

Missing number: ____

Fact family:

____

____

____

____

3. Fact Triangle: top 54; bottom: 6, ___?___; *, /

Missing number: ____

Fact family:

____

____

____

____

The students in Ms. Gonzalez's class estimated the time they spent on homework each week. The tally chart shows the data collected.

4. Construct a line plot for the data.

Student Data on Homework Time

Number of Students

1 2 3 4 5 6

Hours

Number of Hours per Week Spent Studying	Number of Students
1	/
2	//
3	~~////~~ /
4	~~////~~ ~~////~~
5	///
6	//

5. Estimate the amount of time you spend on homework each week. __________

Name ______ Date ______ Time ______

Practice Set 21 *continued*

SRB 17 20 42–44 175

COMPUTATION PRACTICE

Solve.

6. 80 ÷ 8 = ______

7. 30 * 80 = ______

8. 800 = 8 * ______

9. 4 * 400 = ______

10. 30 * ______ = 1,500

11. ______ ÷ 1,000 = 8

12. 1,400 ÷ 700 = ______

13. 28 * ______ = 560

14. ______ ÷ 70 = 70

15. 6 * 30 = ______

16. 4,500 ÷ ______ = 5

17. 9 * 90 = ______

18. How much money, without tax, will you need to buy 3 boxes of crackers that cost $1.59 each? ______

19. How many dollars are in 18 five-dollar bills? ______

20. If 1 block is 200 meters long, how far will you run in 7 blocks?

______

Write a fraction to represent each picture.

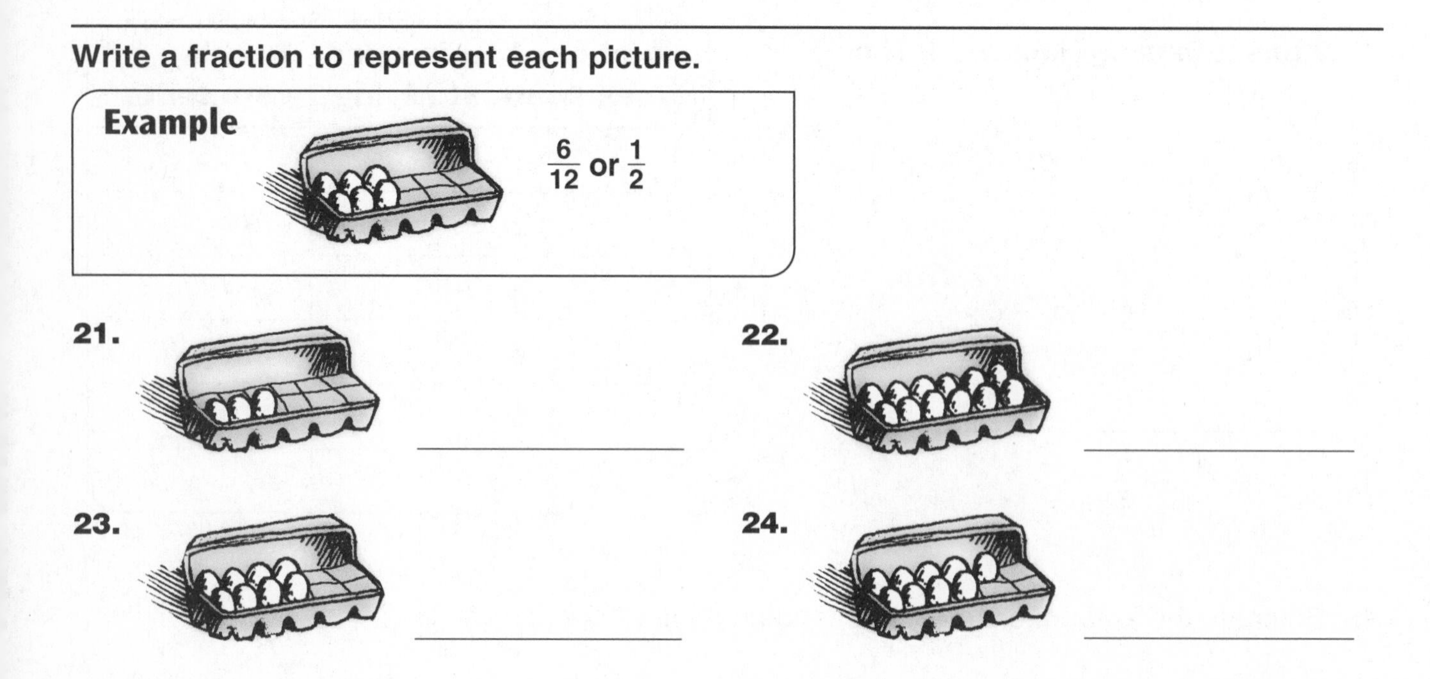

Example

$\frac{6}{12}$ or $\frac{1}{2}$

21. ______

22. ______

23. ______

24. ______

Name Date Time

Practice Set 22

SRB 16–22 41 42

Find the missing number for each Fact Triangle. Then write the fact family for that triangle.

1. Fact Triangle: 45 (top), 5, ? ; *, ÷

Missing number: ____

Fact family:

______

______

______

______

2. Fact Triangle: 36 (top), ?, 12 ; *, ÷

Missing number: ____

Fact family:

______

______

______

______

3. Fact Triangle: ? (top), 7, 6 ; *, ÷

Missing number: ____

Fact family:

______

______

______

______

4. Fact Triangle: ? (top), 4, 15 ; *, ÷

Missing number: ____

Fact family:

______

______

______

______

Write the fraction for the shaded part of the figure.

5. ______

6. ______

7. ______

8. ______

9. ______

10. ______

Name Date Time

Practice Set 23

SRB 128–130, 137

Write your answers in the space provided.

1. The Red Sea covers an area of about 174,000 square miles. Lake Michigan covers an area of about 22,300 square miles. About how many miles greater is the area of the Red Sea than the area of Lake Michigan?

   Number model: ________________ **About** ________ **miles**

2. The deepest part of Lake Michigan is 923 feet. The deepest part of the Red Sea is 8,200 feet. About how many feet deeper is the Red Sea?

   Number model: ________________ **About** ________ **feet**

3. About how many more times deep is the Red Sea than Lake Michigan?

   Number model: ________________ **About** ________ **more times**

4. The Red Sea contains about 1,100 types of fish. Lake Michigan has about 100 types of fish. How many more types of fish are in the Red Sea?

   Number model: ________________ **About** ________ **types**

5. **Writing/Reasoning** Given the information in Problem 4, can you tell if there are more fish in the Red Sea than in Lake Michigan? Explain your answer.

   ____

   ____

Write your answers below.

6. 6 cups = _____ pints
7. 10 quarts = _____ gallons
8. 36 feet = _____ yards
9. 128 inches = _____ feet

COMPUTATION PRACTICE **Add.**

10. $4{,}975 + 1{,}265$
11. $2{,}390 + 1{,}783$
12. $1{,}640 + 9{,}870$
13. $765 + 1{,}832$

Name Date Time

Practice Set 24

SRB 10–19 175

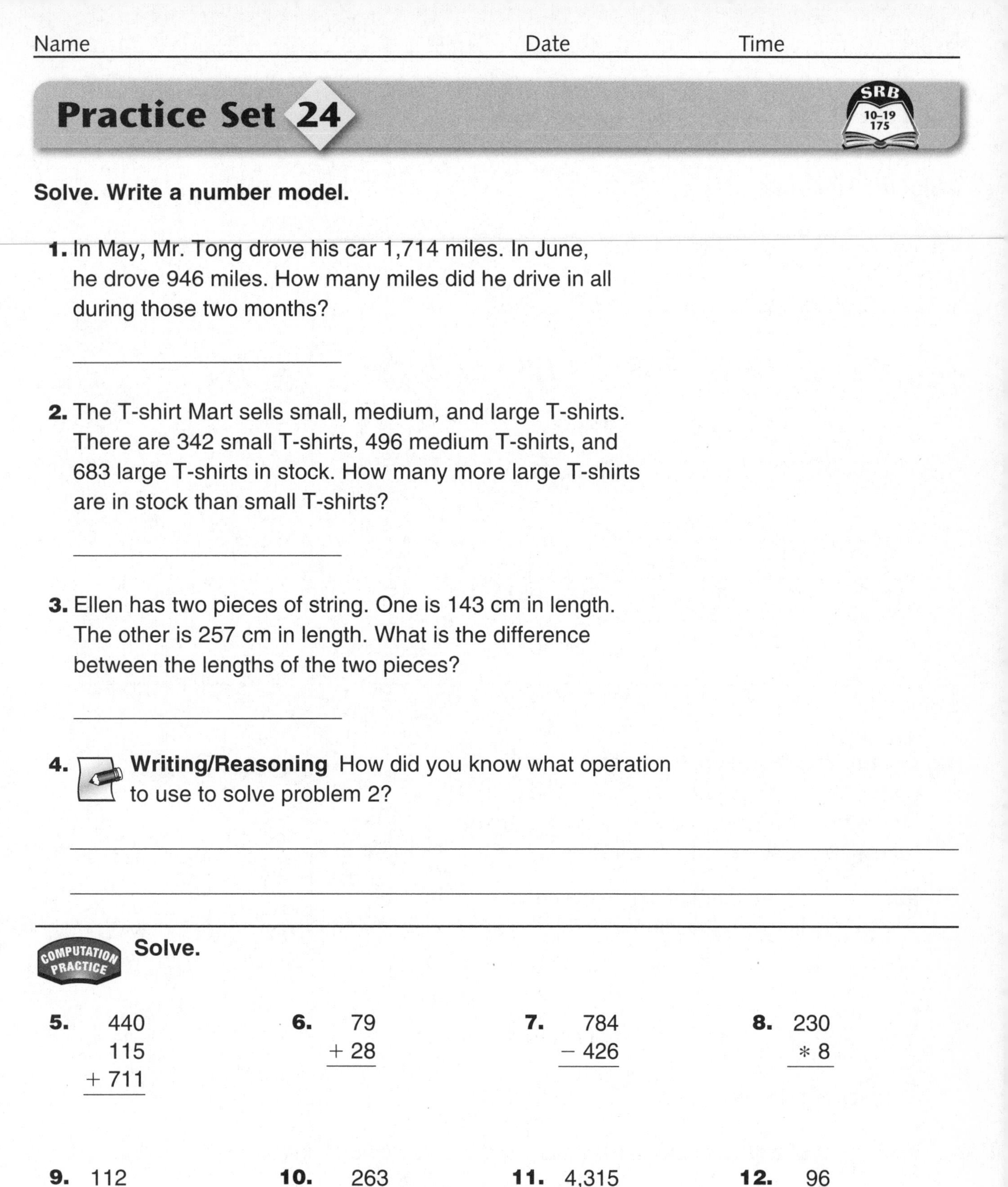

Solve. Write a number model.

1. In May, Mr. Tong drove his car 1,714 miles. In June, he drove 946 miles. How many miles did he drive in all during those two months?

____

2. The T-shirt Mart sells small, medium, and large T-shirts. There are 342 small T-shirts, 496 medium T-shirts, and 683 large T-shirts in stock. How many more large T-shirts are in stock than small T-shirts?

____

3. Ellen has two pieces of string. One is 143 cm in length. The other is 257 cm in length. What is the difference between the lengths of the two pieces?

____

4. **Writing/Reasoning** How did you know what operation to use to solve problem 2?

____

____

COMPUTATION PRACTICE

Solve.

5. 440 + 115 + 711

6. 79 + 28

7. 784 − 426

8. 230 * 8

9. 112 * 9

10. 263 + 357 + 198

11. 4,315 − 78

12. 96 * 3

Name Date Time

Practice Set 24 *continued*

SRB 26 73 76

Write the amounts.

13. (Q) (Q) (Q) (Q) (Q) (D) (D) (N) (N) (N) (P) (P) (P)

____

14. [$1] [$1] (Q) (Q) (Q) (D) (D) (D) (D) (N) (P) (P)

____

15. [$5] [$5] [$5] [$1] (Q) (N) (N) (N)

____

16. [$100] [$20] [$20] [$5] [$1] [$1] [$1] (Q)

____

Solve.

17. Ms. Brown's class kept track of the number of hours they spent reading each day. The graph shows the number of hours the students spent reading Monday through Wednesday.

Hours Spent Reading

Number of Hours

5
4
3
2
1

Mon. Tues. Wed.

a. How many more hours did they read on Monday than on Tuesday?

____

b. What is the average number of hours they spent reading in a day? ______________

c. 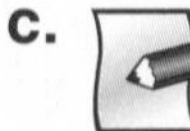 **Writing/Reasoning** How many total hours do you think they would read Monday through Friday? Explain your answer.

____

____

Name Date Time

Practice Set 25

SRB 9 42–44 148

For each number model, write *T* if it is true, *F* if it is false, or *?* if you can't tell.

1. $8 * 9 = 76$ ________

2. $5 + 9 < 20$ ________

3. $4 = 64 / 8$ ________

4. $26 + 19 = 7$ ________

5. $450 - 119$ ________

6. $18 < 15 + 6$ ________

7. $70 - 21 = 49$ ________

8. $9 * 4 > 36$ ________

Write a number for each picture below.
Use 0 or $\frac{0}{4}$, $\frac{1}{4}$, $\frac{1}{2}$ or $\frac{2}{4}$, $\frac{3}{4}$, and 1 or $\frac{4}{4}$.

9. FUEL E F

10. FUEL E F

11. FUEL E F

COMPUTATION PRACTICE

Solve.

12. $47 * 6$

13. $63 * 3$

14. $214 * 5$

15. $703 * 7$

16. $115 * 8$

17. $98 * 4$

18. $335 * 2$

19. $750 * 6$

List the next four multiples.

20. 7, ______, ______, ______, ______

21. 3, ______, ______, ______, ______

22. 4, ______, ______, ______, ______

23. 9, ______, ______, ______, ______

Name Date Time

Practice Set 26

SRB 150 151 175

Rewrite the number models with parentheses to make them correct.

1. $6 * 8 - 3 = 45$

____

2. $22 = 8 + 3 * 2$

____

3. $33 - 15 - 6 = 24$

____

4. $54 - 10 + 8 = 52$

____

5. $3 * 8 + 2 * 11 = 46$

____

6. $30 = 4 * 6 + 6$

____

7. $2 * 2 + 7 * 8 = 60$

____

8. $489 = 5 * 25 + 75 - 11$

____

---

9. The bases on a baseball diamond are placed exactly 90 feet apart.

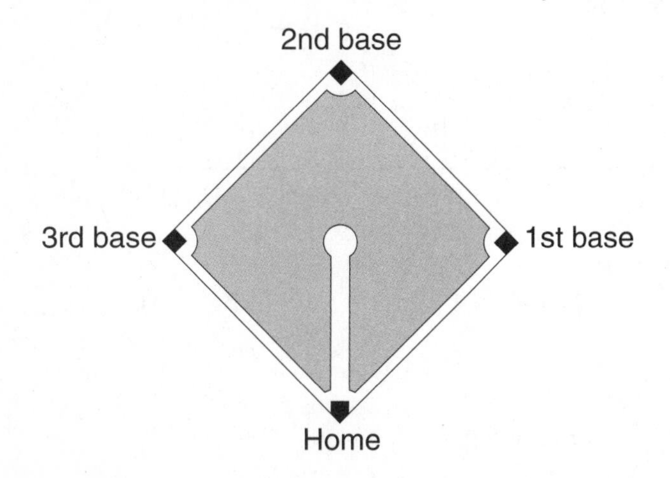

a. If a batter hits a home run, how many feet does she run? _______________

b. If there are runners on first and third when a batter hits a home run, what is the total distance all three players run? _______________

10. Write the following number in digits:
eight thousand, four hundred twenty-one. _______________

11. Write the words for 1,603.

____

Name Date Time

Practice Set 27

Find the solution of each open sentence below. Write a number sentence with the solution in place of the variable.

1. $x + 8 = 35$ _______________

2. $4t = 24$ _______________

3. $140 + 3 = y$ _______________

4. $\frac{s}{7} = 7$ _______________

5. $m - 60 = 200$ _______________

6. $68 + r = 80$ _______________

7. $6x = 42$ _______________

8. $\frac{70}{n} = 10$ _______________

Rewrite the number sentences with parentheses to make them correct.

9. 204 = 7 * 20 + 75 – 11 _______________

10. 7 * 9 – 4 = 35 _______________

11. 42 = 3 + 3 * 7 _______________

12. 31 – 15 – 6 = 10 _______________

13. 54 – 10 + 8 = 52 _______________

14. 7 * 8 + 3 * 11 = 89 _______________

COMPUTATION PRACTICE **Solve.**

15. 212 * 20

16. 785 – 76

17. 867 – 74

18. 76 * 0

19. 7,210 + 9,188

20. 600 – 599

21. (60 + 80) * 4 = _______________

22. 39 – (3 * 4) = _______________

23. 620 + 150 + 220 = _______________

24. (1,800 ÷ 60) * 5 = _______________

Name Date Time

Practice Set 27 *continued*

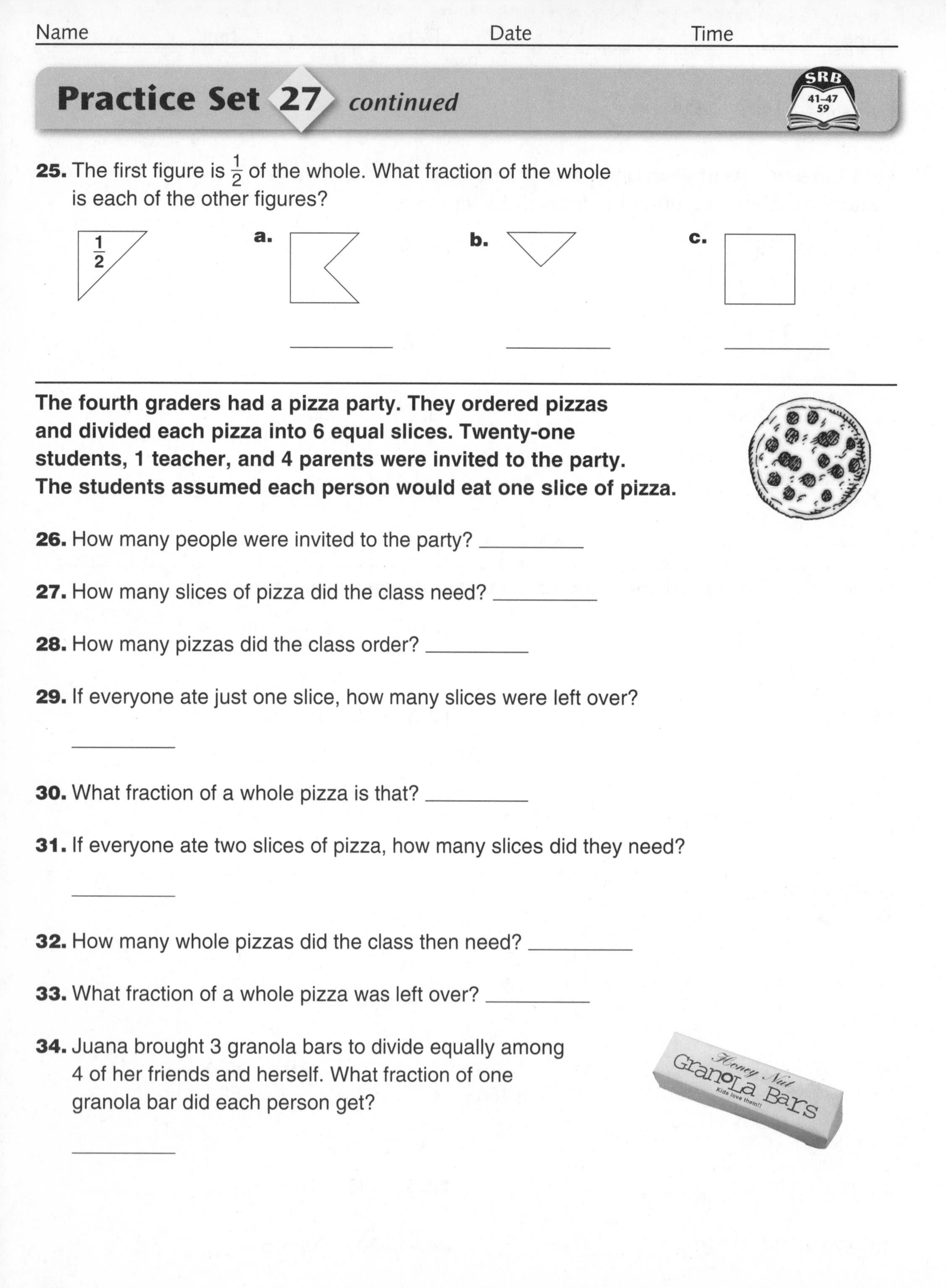

25. The first figure is $\frac{1}{2}$ of the whole. What fraction of the whole is each of the other figures?

a. ______

b. ______

c. ______

The fourth graders had a pizza party. They ordered pizzas and divided each pizza into 6 equal slices. Twenty-one students, 1 teacher, and 4 parents were invited to the party. The students assumed each person would eat one slice of pizza.

26. How many people were invited to the party? ______

27. How many slices of pizza did the class need? ______

28. How many pizzas did the class order? ______

29. If everyone ate just one slice, how many slices were left over?

______

30. What fraction of a whole pizza is that? ______

31. If everyone ate two slices of pizza, how many slices did they need?

______

32. How many whole pizzas did the class then need? ______

33. What fraction of a whole pizza was left over? ______

34. Juana brought 3 granola bars to divide equally among 4 of her friends and herself. What fraction of one granola bar did each person get?

______

Name Date Time

Practice Set 28

SRB 30–33 175

Complete the number patterns.

1. ______, ______, 0, 8, ______, 24, ______

2. 2, 24, ______, ______, 90, ______, ______

3. ______, 11, ______, 33, ______, ______, 66

Write the numbers in order from least to greatest.

4. 1.2, 0.2, 2.10, 2.2

____

5. 0.23, 1.2, 0.04, 5.1

____

6. 4.01, 1.4, 2.14, 1.41

____

7. 9.5, 1.95, 19.5, 0.59

____

8. 0.3, 3.3, 0.03, 3.03

____

9. 5.20, 5.12, 5.02, 5.21

____

COMPUTATION PRACTICE

Solve.

10. 420 − ________ = 81

11. 712 + 517 = ________

12. 160 + 348 = ________

13. 490 − 170 = ________

14. 2,216 − 1,804 = ________

15. 90 = 8,100 ÷ ________

16. 387 − 36

17. 673 − 615

18. 57 * 7

19. 7,619 + 3,250

20. 980 * 50

21. 427 + 561 + 711

22. Alvin wants to wear a different pair of socks for each of the 14 days he will be on vacation. How many socks does he need to pack? ________

Name Date Time

Practice Set 28 *continued*

SRB 4 160–161

Complete the frames-and-arrows problems.

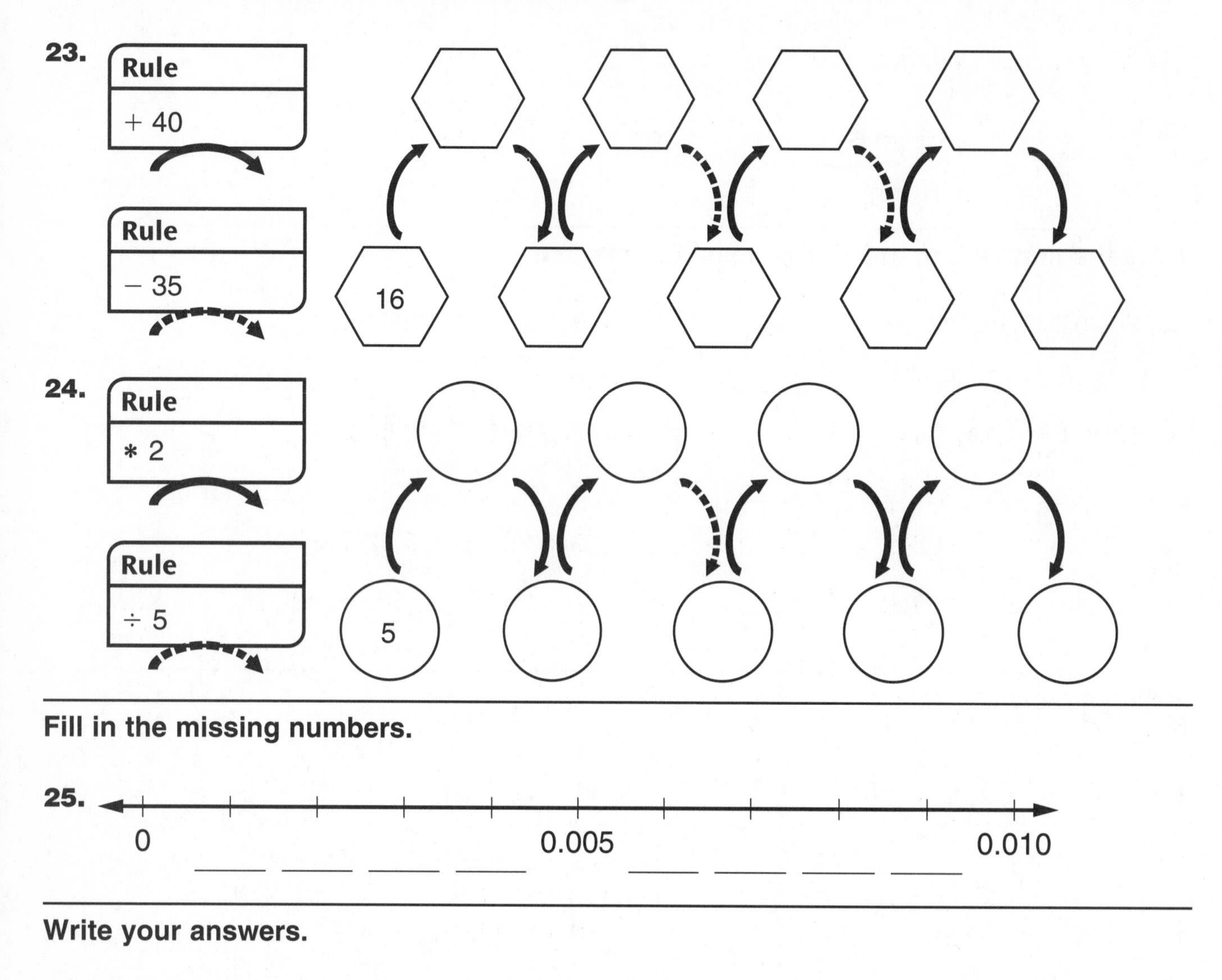

Fill in the missing numbers.

25. 0 ____ ____ ____ ____ 0.005 ____ ____ ____ ____ 0.010

Write your answers.

26. Write the following in digits: five thousandths.

____

27. Write the number words for 0.03.

____

28. Write the number words for 0.76.

____

Use with or after Lesson 4•1.

Name Date Time

Practice Set 29

Ms. Gomez is a food service director. She bought the fruits and vegetables shown in the table for a PTA dinner. Use estimation to answer the following questions.

Item	Amount
apples	15.2 kg
oranges	19.7 kg
watermelon	17.3 kg
celery	12.1 kg
potatoes	33.8 kg
tomatoes	6.9 kg

1. Which item did she buy the most of in terms of weight?

____

2. Which item did she buy the least of in terms of weight?

____

3. About how many more kilograms of oranges did she buy than kilograms of apples?

____

4. Mr. Hong's class took a survey to find out how many brothers and sisters each student has. Use the information in the tally chart to complete the line plot.

Number of Siblings	
0	///
1	~~////~~ ////
2	~~////~~
3	/
4	///
5	
6	/

Number of Students

0 1 2 3 4 5 6

Number of Siblings

Write a whole-number estimate.

5. 3.7 + 9.9 **Estimate:** __________

6. 10.6 − 6.8 **Estimate:** __________

7. 3.7 + 4.5 **Estimate:** __________

8. 24.8 − 7.8 **Estimate:** __________

Name Date Time

Practice Set 29 *continued*

SRB 160 161

Complete the frames-and-arrows problems.

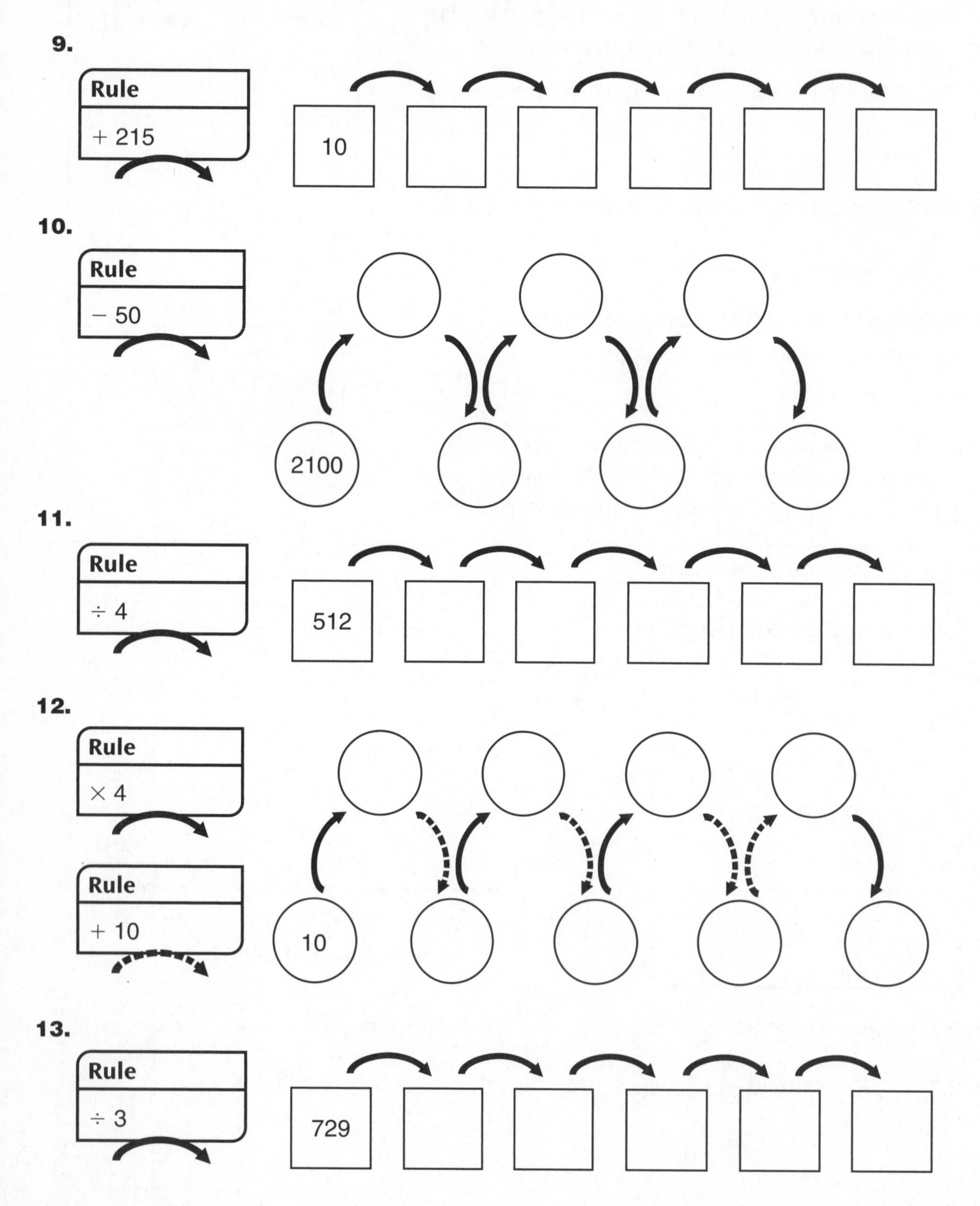

Use with or after Lesson 4•4.

Name Date Time

Practice Set 30

SRB 34–37 128 162

COMPUTATION PRACTICE

Add or subtract.

1. 15.3 + 12.9 = ________

2. 5 − 1.6 = ________

3. 11.3 − 6 = ________

4. 8.24 + 3.9 = ________

5. 2.09 + 2.2 = ________

6. 1.9 − 0.41 = ________

7. 5.15 − 3.67 = ________

8. 0.69 + 0.75 = ________

9. 18.5 + 3.9 = ________

10. 1.7 − 1.34 = ________

Measure to the nearest inch and the nearest centimeter.

11. ______________________________

________ inches ________ centimeters

12. ______________________

________ inches ________ centimeters

13. ____

________ inches ________ centimeters

Complete the "What's My Rule?" tables and rule box.

14.

Rule
out = in * 200

in	out
7	1,400
9	
12	
14	
35	

15.

Rule

in	out
7	$14\frac{1}{2}$
10	$17\frac{1}{2}$
	20
$13\frac{1}{2}$	
$22\frac{1}{2}$	30

Name _______________ Date _______ Time _______

Practice Set 30 *continued*

SRB 22–24 175

Answer the following questions.

16. Four people are going to share $68 equally.

a. How many $10 bills does each person get? ______________

b. How many dollars are left to share? ______________

c. From the money that remains, how many $1 bills does each person get? ______________

d. What is the total amount of money each person gets? ______________

e. Number model: 4 * 17 = ☐

17. Seven people are going to share $112 equally.

a. How many $100 bills does each person get? ______________

b. From the money that remains, how many $10 bills does each person get? ______________

c. How many dollars are left to share? ______________

d. From the money that remains, how many $1 bills does each person get? ______________

e. What is the total amount of money each person gets? ______________

f. Write a number model for this problem. ______________

18. Six people are going to share $681 equally.

a. How many $100 bills does each person get? ______________

b. How many dollars are left to share? ______________

c. From the money that remains, how many $10 bills does each person get? ______________

d. How many dollars are left to share? ______________

e. From the money that remains, how many $1 bills does each person get? ______________

f. How many dollars are left over? ______________

Name Date Time

Practice Set 31

SRB 34-37 149 162 163

COMPUTATION PRACTICE

Solve.

1. \$17.03 − \$3.85 = ____________

2. \$8.22 + \$30 = ____________

3. \$6.05 + \$2.80 = ____________

4. \$16.23 − \$9.66 = ____________

5. \$14.85 − \$12.60 = ____________

6. \$5.99 + \$2.49 = ____________

7. \$20 + \$16.56 = ____________

8. \$15.68 − \$5.97 = ____________

9. \$7.48 + \$21.70 = ____________

10. \$50 − \$19.79 = ____________

Complete the "What's My Rule?" tables. Write the missing rules.

11.

Rule
out = in * 12

in	out
3	36
5	
7	84
11	
15	

12.

Rule

in	out
8	24
14	30
	44
35	
43	59

Make a name-collection box for each number listed below. Use as many different numbers and operations as you can.

Example

419
$(\frac{1}{5}$ * 2,500) − (9 * 9)
(205 * 2) + 9
838 ÷ 2
(500 − 85) + 4

13.

380

14.

176

15.

4,218

Name Date Time

Practice Set 32

SRB 61 148 152–154

Complete the table. The ☐ is ONE.

Base-10 Blocks	Fraction Notation	Decimal Notation
1. 	$\frac{258}{1,000}$	
2. 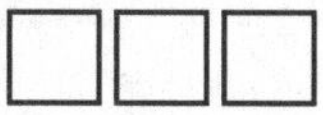	$\frac{3}{10}$	
3. 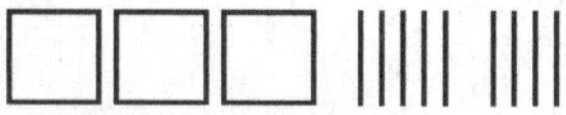		0.39
4. 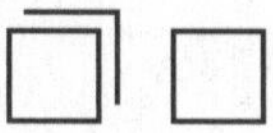		1.10
5.	$\frac{90}{1,000}$	0.090
6. 		0.008
7. 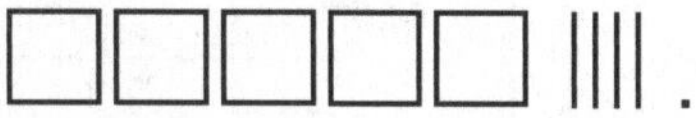	$\frac{541}{1,000}$	

Name _______________ Date _______ Time _______

Practice Set 33

SRB 126 129 148 149

Measure each line segment to the nearest centimeter.
Record the measurements in centimeters and meters.

1. ______ cm; ______ m

2. ______ cm; ______ m

3. ______ cm; ______ m

4. ______ cm; ______ m

5. ______ cm; ______ m

For each number sentence, write *T* if it is true or *F* if it is false.

6. $92 - 40 = 42$ __________

7. $80 = 18 + 62$ __________

8. $9 * 7 = 67$ __________

9. $6 > \frac{72}{9}$ __________

10. $500 + 80 < 550$ __________

11. $448 - 15 > 400$ __________

Use digits to write the following numbers.

12. two hundred sixty thousand, four hundred fifty-three

____

13. two hundred eighty-six and thirty-eight hundredths

____

14. three hundred fourteen thousand, six hundred ninety-one

____

15. one million, seventy-four thousand, nine hundred sixty-eight

____

16. six million, seven hundred nine thousand, eight hundred forty-five

____

Name Date Time

Practice Set 33 *continued*

SRB 16–24 175

COMPUTATION PRACTICE **Solve.**

17. 180 + 6 = ______________

18. 24 * 6 = ______________

19. 810 = 90 * ______________

20. 40 * 70 = ______________

21. 80 * ______________ = 3,200

22. ______________ ÷ 100 = 6

23. 8,400 ÷ 700 = ______________

24. 36 * ______________ = 72

25. ______________ ÷ 5 = 70

26. 9 * 200 = ______________

27. 5,400 ÷ ______________ = 9

28. 12 * 120 = ______________

Write your answers.

29. How much money, without tax, will you need to buy 4 greeting cards that cost $3.25 each?

____

30. **Writing/Reasoning** Explain how you found the answer to Problem 29.

____

____

____

31. How many dollars are in 22 five-dollar bills? ______________

32. If 1 block is 200 meters long, how far will you run in 23 blocks?

____

33. There are about 8 blocks in one mile. How many blocks are in 5 miles? ______________

34. There are 5,280 feet in one mile. How many yards are in one mile?

(Reminder: 1 yd = 3 ft) ______________

Name Date Time

Practice Set 34

SRB 10–15 128 129

Complete.

1. 500 cm = ________ m
2. ________ cm = 8.1 m
3. ________ mm = 72 cm
4. 150 cm = ________ m
5. 0.35 m = ________ cm
6. 63 cm = ________ m
7. ________ mm = 9.8 cm
8. ________ m = 375 cm
9. **Writing/Reasoning** Explain how you found the answer to Problem 8.

____

____

COMPUTATION PRACTICE

Solve.

10. 1,800 − 927

11. 3,684 − 485

12. 3,164 + 5,791

13. 8,261 − 3,540

14. 600 − 31

15. 475 + 250

16. 1,834 + 8,365

17. 469 − 70

18. 1,200 − 30

19. 2,444 − 382

20. 729 + 682

21. 1,356 − 1,172

22. 4,321 − 1,234

23. 500 − 42

24. 1,300 − 485

Name Date Time

Practice Set 35

SRB 4 9 26–31 127–129

1. Use the clues to complete the place-value puzzle.

- Add 43 and 23. Divide by 11 and write the result in the thousandths place.
- Triple the number in the thousandths place and divide by 2. Write the result in the hundredths place.
- Multiply 8 * 9. Subtract 68. Write the result in the ones place.
- Subtract the number in the hundredths place from 57 and divide by 6. Write the result in the hundreds place.
- Divide 36 by the number in the ones place. Write the result in the tens place.
- Subtract the number in the tens place from the number in the hundredths place. Write the result in the tenths place.

100s	10s	1s	.	0.1s	0.01s	0.001s

2. Measure the pencil to the nearest millimeter.

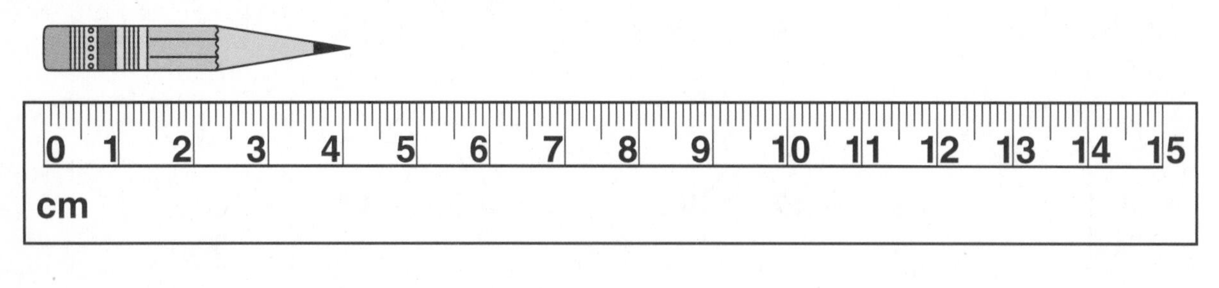

______________ mm

3. **Writing/Reasoning** The picture of the bumblebee is 2 times its actual size. This picture is 4 cm long. What is the actual length of the bumblebee? Explain how you found your answer.

____

____

____

Use with or after Lesson 4•10.

Name Date Time

Practice Set 35 *continued*

SRB 150 162 163

Complete.

4. 4.3 cm = ______ mm
5. 30 mm = ______ cm
6. 6 mm = ______ cm
7. 4.3 cm = ______ mm

Complete the "What's My Rule?" tables. Write the missing rule.

8.

Rule

in	out
10	18.25
12.5	
	22.75
54.10	
88.0	96.25

9.

Rule
out = in ÷ 3

in	out
600	200
900	
1,200	
1,500	
1,800	

Rewrite the number sentences with parentheses to make them correct.

10. $26 \div 2 - 7 = 6$ ______
11. $41.2 = 7 * 6 - 0.8$ ______
12. $130 - 15 - 60 = 55$ ______
13. $118 = 2 * 55.7 + 3.3$ ______
14. $10 * 2.1 + 5 * 12.2 = 82$ ______
15. $30 = 6 * 20 - 90$ ______
16. $11 * 12 + 7 - 4 = 165$ ______
17. $99 = 11 * 50 - 41$ ______
18. $50 + 300 \div 5 = 70$ ______
19. $200 * 2.6 - 1.1 + 5 = 305$ ______

Name Date Time

Practice Set 36

SRB 16–19 160 161

COMPUTATION PRACTICE **Solve.**

1. 4 * 80 = ________ **2.** 60 * 50 = ________ **3.** 20 * 7 = ________

4. 5 * 40 = ________ **5.** 70 * 60 = ________ **6.** 30 * 9 = ________

7. 80 * 20 = ________ **8.** 40 * 80 = ________ **9.** 40 * 900 = ________

A bag contains 5 blue marbles, 6 red marbles, 2 green marbles, 1 orange marble, and 7 purple marbles.

10. What is the chance of picking a red marble without looking?

________ out of ________

11. What is the chance of picking a green marble or an orange marble without looking?

________ out of ________

12. What is the chance of picking a blue marble or a purple marble without looking?

________ out of ________

13. What is the chance of picking a black marble without looking?

________ out of ________

COMPUTATION PRACTICE **Solve.**

14. (3 + 4) * (5 + 2) = ________ **15.** (2 + 4) * (3 + 5) = ________

16. (5 + 4) * (2 + 3) = ________ **17.** (3 + 2) * (4 + 5) = ________

18. 0.4 + 0.05 ________ **19.** 0.3 + 0.23 = ________

20. 0.45 − 0.04 = ________ **21.** 0.88 − 0.08 = ________

Name Date Time

Practice Set 36 *continued*

Kloe asked 35 teachers and students the month of their birthday. She kept a tally chart of their responses.

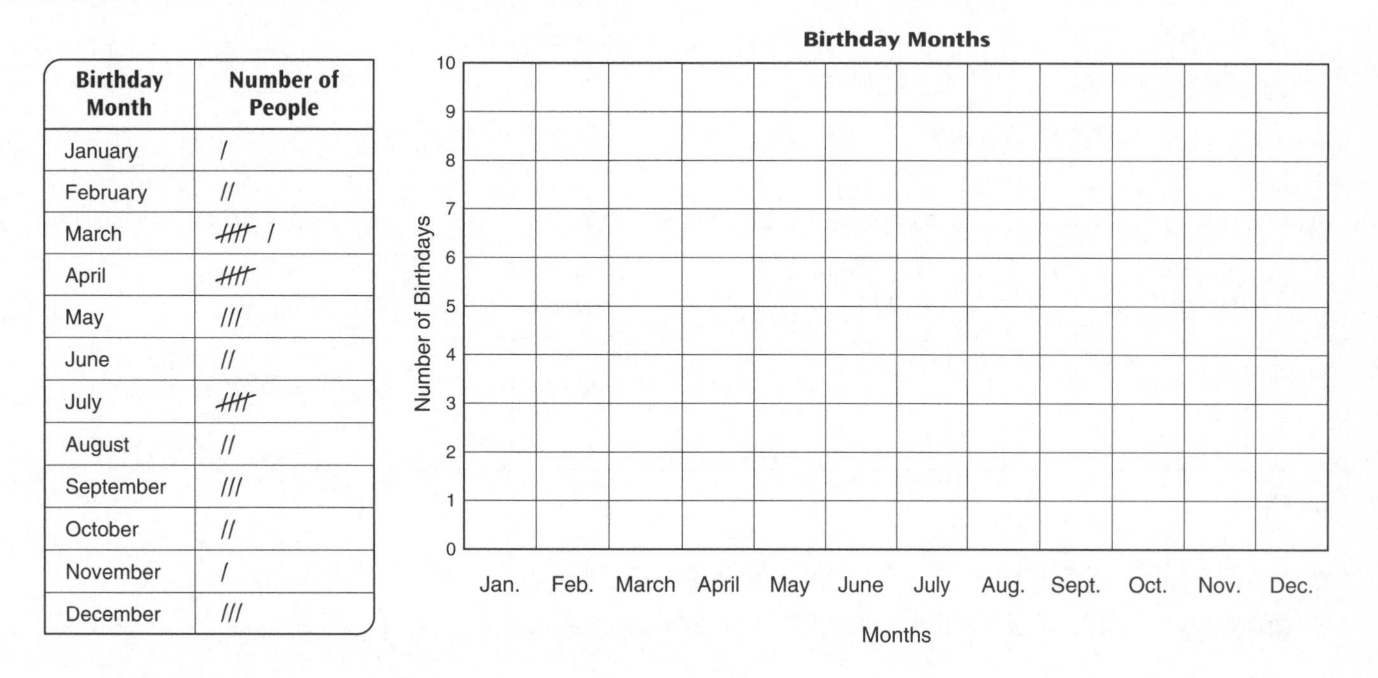

Birthday Month	Number of People
January	/
February	//
March	卌 /
April	卌
May	///
June	//
July	卌
August	//
September	///
October	//
November	/
December	///

22. Use the information in the tally chart to complete the bar graph.

23. Which month has the most birthdays? __________

24. Which months have the least birthdays? __________

25. What is the mean (average) number of birthdays each month?

____

26. What is the median? __________

27. What is the mode? __________

28. What is the range? __________

Name ______ Date ______ Time ______

Practice Set 37

SRB 10–22 180–183

Estimate the sums.

1.
```
  2,860
  1,452
+ 9,024
```

2.
```
  1,010
  1,345
+ 6,813
```

3.
```
  5,272
  2,470
+ 1,391
```

4.
```
  7,280
  6,109
+ 6,109
```

5.
```
  3,554
  7,490
  6,708
+ 2,143
```

6.
```
  9,656
  1,970
  6,213
+ 3,160
```

7.
```
  2,843
  7,154
  1,041
  2,630
+ 1,274
```

8.
```
  6,590
  4,483
  3,217
  1,670
+ 2,532
```

Solve.

9. You went to the store with a $10 bill and a $5 bill. Your groceries cost $12.36. How much change should you get? ______

10. $\frac{20}{4}$ = ______ **11.** $\frac{36}{9}$ = ______ **12.** 7 * 7 = ______ **13.** 6 * 6 = ______

14. 9 * 80 = ______ **15.** 12 + 13 = ______ **16.** 41 − 25 = ______ **17.** 7 * 8 = ______

18. 5 * 6 = ______ **19.** $\frac{12}{12}$ = ______ **20.** $\frac{40}{10}$ = ______ **21.** $\frac{48}{6}$ = ______

Use digits to write the following numbers.

22. forty-five thousand, three hundred ninety-two

______

23. four hundred fifty-nine thousand, seven hundred three

______

24. six million, four thousand, six hundred two

______

Name Date Time

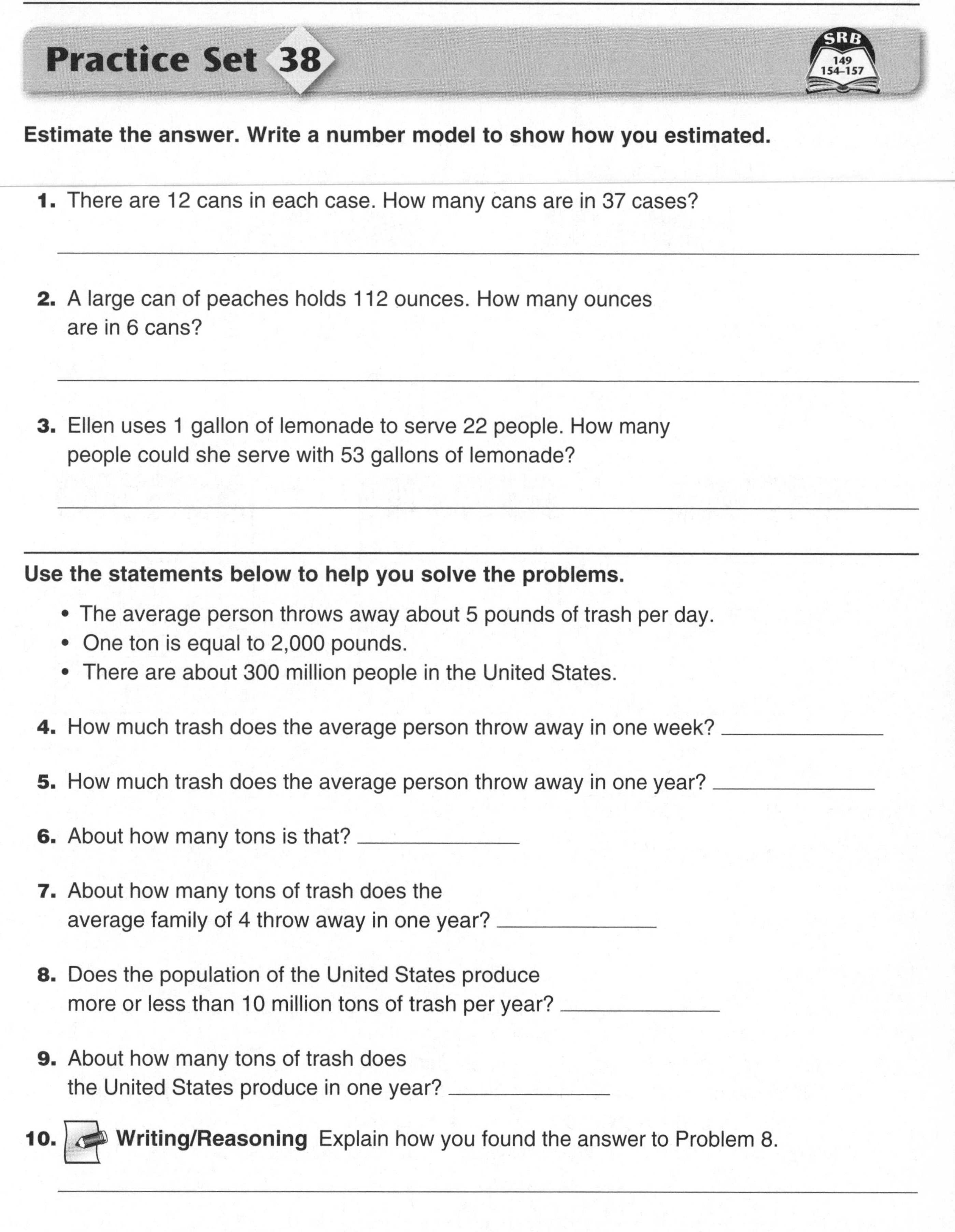

Practice Set 38

SRB 149 154–157

Estimate the answer. Write a number model to show how you estimated.

1. There are 12 cans in each case. How many cans are in 37 cases?

____

2. A large can of peaches holds 112 ounces. How many ounces are in 6 cans?

____

3. Ellen uses 1 gallon of lemonade to serve 22 people. How many people could she serve with 53 gallons of lemonade?

____

Use the statements below to help you solve the problems.

- The average person throws away about 5 pounds of trash per day.
- One ton is equal to 2,000 pounds.
- There are about 300 million people in the United States.

4. How much trash does the average person throw away in one week? _______________

5. How much trash does the average person throw away in one year? _______________

6. About how many tons is that? _______________

7. About how many tons of trash does the average family of 4 throw away in one year? _______________

8. Does the population of the United States produce more or less than 10 million tons of trash per year? _______________

9. About how many tons of trash does the United States produce in one year? _______________

10. **Writing/Reasoning** Explain how you found the answer to Problem 8.

____

____

Name Date Time

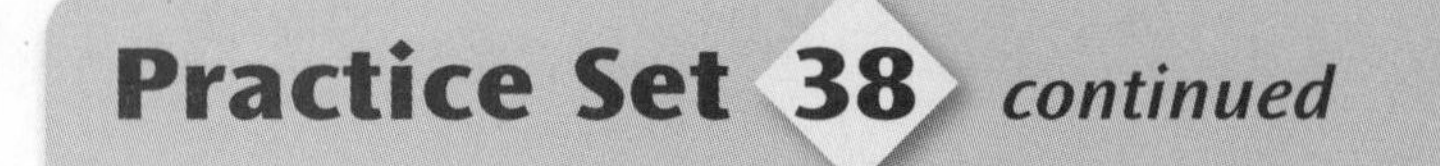

Practice Set 38 *continued*

Make a name-collection box for each number listed below.
Use as many different numbers and operations as you can.

Example

24
XXIV
$48 \div 2$
$29 - 5$
12×2

11.

15

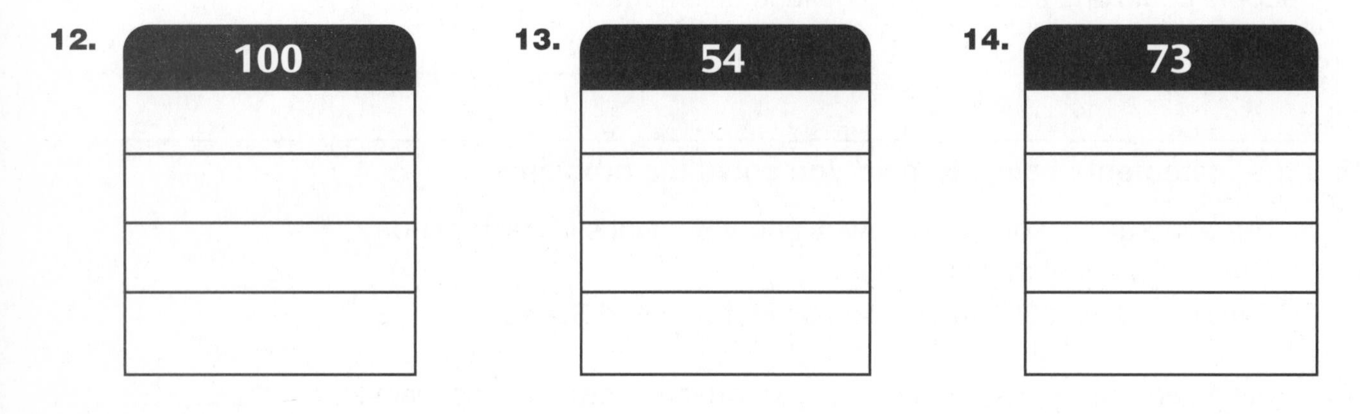

True or False?

Example

$12 + 15 = 25$ *False*

$4 * (3 + 1) = 16$ *True*

15. $7 * 9 = 54$ ______________

16. $3 * (4 + 5) = 27$ ______________

17. $5 * 6 = 40$ ______________

18. $(99 + 13) = 103$ ______________

19. Use the clues to complete the place-value puzzle.

- Divide 18 by 6. Write the result in the ones place.
- Double the number in the ones place. Divide by 3. Write the result in the tens place.
- Write the result of $8 * 5$ divided by 10 in the thousands place.
- Multiply 7 by 2. Subtract 7. Write the result in the hundreds place.

1,000s	100s	10s	1s

Name Date Time

Practice Set 39

SRB 4 16–18 180–184

COMPUTATION PRACTICE

Multiply using the partial-product method. Show your work in the grid.

1. 82 * 9 = ________ **2.** 135 * 6 = ________ **3.** 545 * 3 = ________

Estimate whether the answer will be in the tens, hundreds, or thousands. Write a number model to show how you estimated. Circle the box that shows your estimate.

4. Jabir collected 214 pennies each week for 6 weeks. How many pennies did he collect?

Number model: ________ **Exact answer:** ________

10s	100s	1,000s

5. Kaya's piano lesson lasts 130 minutes. If she has 4 lessons this week, how many minutes will Kaya spend at piano lessons?

Number model: ________ **Exact answer:** ________

10s	100s	1,000s

Use digits to write the numbers.

6. one hundred and seventy-five thousand, three hundred forty-two ________

7. twenty-seven and eight tenths ________

8. three million, eighty-two thousand, six hundred two ________

Name Date Time

Practice Set 40

SRB 16–24 75

COMPUTATION PRACTICE

Use partial products to multiply.

1. 47 * 6

2. 89 * 5

3. 27 * 13

4. 36 * 46

5. 90 * 19

6. 23 * 62

7. 159 * 41

8. 613 * 18

COMPUTATION PRACTICE

Solve.

9. 23 * 4 = ________

10. 18 ÷ 1 = ________

11. 54 + 36 = ________

12. 9 * 35 = ________

13. $\frac{180}{10}$ = ________

14. 78 − 23 = ________

15. 50 * 60 = ________

16. $\frac{162}{6}$ = ________

17. 48 − 12 = ________

Examine the data sets. Find the mean of each.

Example 7 6 5 9 8

Step 1 Find the total of the numbers in the data set. 7 + 6 + 5 + 9 + 8 = 35
Step 2 Count how may numbers are in the data set. There are 5 numbers in all.
Step 3 Divide the total by the amount of numbers. $\frac{35}{5}$ = 7 **Mean = 7**

18. 5 8 8 11 7 9

____

19. 12 10 8 14 11

____

20. 3 2 3 1 6 4 2

____

21. 14 15 18 13

____

Name Date Time

Practice Set 40 *continued*

SRB 44 158

Who am I?

22. Clue 1: I am less than 10.
Clue 2: I am an even number.
Clue 3: I am a square number.

I am ______________.

23. Clue 1: I am less than 100.
Clue 2: The sum of my digits is 17.
Clue 3: I am an even number.

I am ______________.

24. Clue 1: I am a number between 100 and 200.
Clue 2: The sum of my digits is 3.
Clue 3: My ones digit is two times my hundreds digit.

I am ______________.

---

25. The first figure is $\frac{3}{4}$ of the whole. What fraction of the whole is each of the other figures?

$\frac{3}{4}$

a. ______________ **b.** ______________ **c.** ______________

26. Write a number for each picture. Use 0 or $\frac{0}{4}$, $\frac{1}{4}$, $\frac{1}{2}$, $\frac{3}{4}$, and $\frac{4}{4}$ or 1.

a. ______________ **b.** ______________

c. ______________ **d.** ______________ **e.** ______________

Name Date Time

Practice Set 41

Use the lattice method to find the following products.

1. 58 * 16 = ________

2. 61 * 107= ________

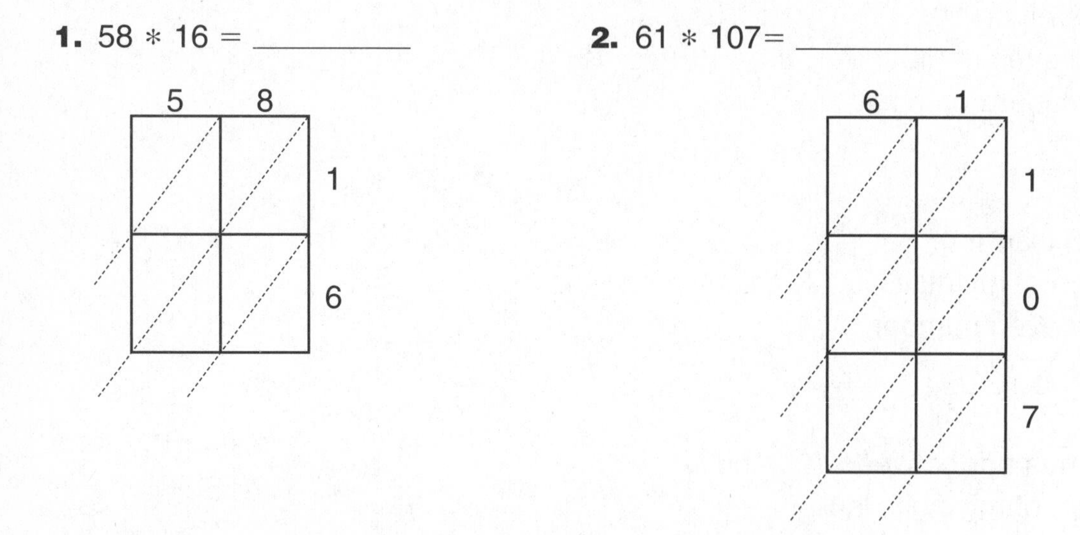

Jamar asked the students in his class to name their favorite sport. He kept a tallychart of what each person said.

3. Use the information in the tally chart to make a line plot. Label the columns and the two axes. Give your graph a title.

Favorite Sport	Number of People
Football	////
Soccer	~~////~~ //
Volleyball	~~////~~ /
Basketball	~~////~~
Track	///

Name Date Time

Practice Set 41 *continued*

SRB 18 19 150 175

COMPUTATION PRACTICE

Solve.

4. $326 * 30$

5. $965 - 86$

6. $541 - 8$

7. $160 + 1{,}400$

8. $6{,}045 + 248$

9. $2{,}289 + 1{,}374$

10. $18 * 11$

11. $4{,}371 + 8{,}148$

12. $890 - 15$

13. $(70 + 15) * 4 =$ ________

14. $67 - (8 * 4) =$ ________

15. $430 + 70 + 145 =$ ________

16. $(\frac{72}{8}) * 3 =$ ________

17. **Writing/Reasoning** Alice recorded the change in temperature every hour. For the past three hours she recorded + 3, −1, and + 2 degrees. Can you tell what the temperature is now? Explain your answer.

____

____

Rewrite the number sentences with parentheses to make them correct.

18. $412 = 70 * 5 + 1 - 8$

19. $6 * 10 - 5 = 55$

20. $81 = 7 + 2 * 9$

21. $39 - 16 - 4 = 19$

22. $44 - 13 + 23 = 8$

23. $8 * 5 + 2 * 18 = 76$

24. $135 = 9 * 9 + 6$

25. $4 * 3 + 7 * 6 = 240$

Name ______________________ Date __________ Time __________

Practice Set 42

SRB 4 26 34–37

Use digits to write the following numbers.

1. two million, seven hundred eighteen thousand, nine hundred twenty

____

2. seven hundred sixty-nine thousand, two hundred thirty-one

____

3. eighteen and nine hundred seventy-eight thousandths

____

Write the number words for the following.

4. 18,564,290 ______________________

5. 48.128 ______________________

6. 5,773,963 ______________________

7. 102,756 ______________________

Write the amounts.

8. $1 Q Q Q Q Q Q Q D P P P P $ __________

9. $5 $1 $1 Q Q Q N N N N $ __________

10. $100 $20 $5 $5 $5 $1 $1 $ __________

11. Q Q Q Q Q Q D N N N P P $ __________

12. $5 $5 $1 $1 Q D N P P $ __________

13. $20 $20 $10 $5 Q Q Q Q D N $ __________

Name Date Time

Practice Set 43

SRB 4 6 10–24

Write the numbers in order from least to greatest.

1. 10,000; 10^5; 10 * 10; 1 thousand

____

2. 1 million; 10^4; 10 [hundreds]; 10 * 10 * 10 * 10 * 10

____

3. 100,000; 10^2; 10 [thousands]; 10 * 10 * 10

____

4. 10 [tenths]; 1 thousand; 10^5; 10 * 10

____

Write the numbers in exponential notation.

5. 1,000 = _____ **6.** 100,000 = _____ **7.** 1,000,000 = _____ **8.** 1 = _____

FACTS PRACTICE

Solve.

9. 20 ÷ 4 = _____ **10.** $\frac{36}{9}$ = _____ **11.** 12 ÷ 12 = _____ **12.** 40 ÷ 10 = _____

13. 48 ÷ 6 = _____ **14.** $\frac{8}{4}$ = _____ **15.** $\frac{0}{4}$ = _____ **16.** 18 ÷ 1 = _____

17. 12 ÷ 4 = _____ **18.** $\frac{24}{8}$ = _____ **19.** 4 ÷ 8 = _____ **20.** $\frac{144}{12}$ = _____

COMPUTATION PRACTICE

Solve.

21. 35 * 4

22. 62 * 3

23. 265 * 4

24. 43 + 18

25. 710 + 136

26. 500 + 4,500

27. 3,249 − 1,933

28. 1,865 − 674

Name ____ Date ____ Time ____

Practice Set 43 *continued*

SRB 175 180–182

Solve the following problems. (The prices include tax.)

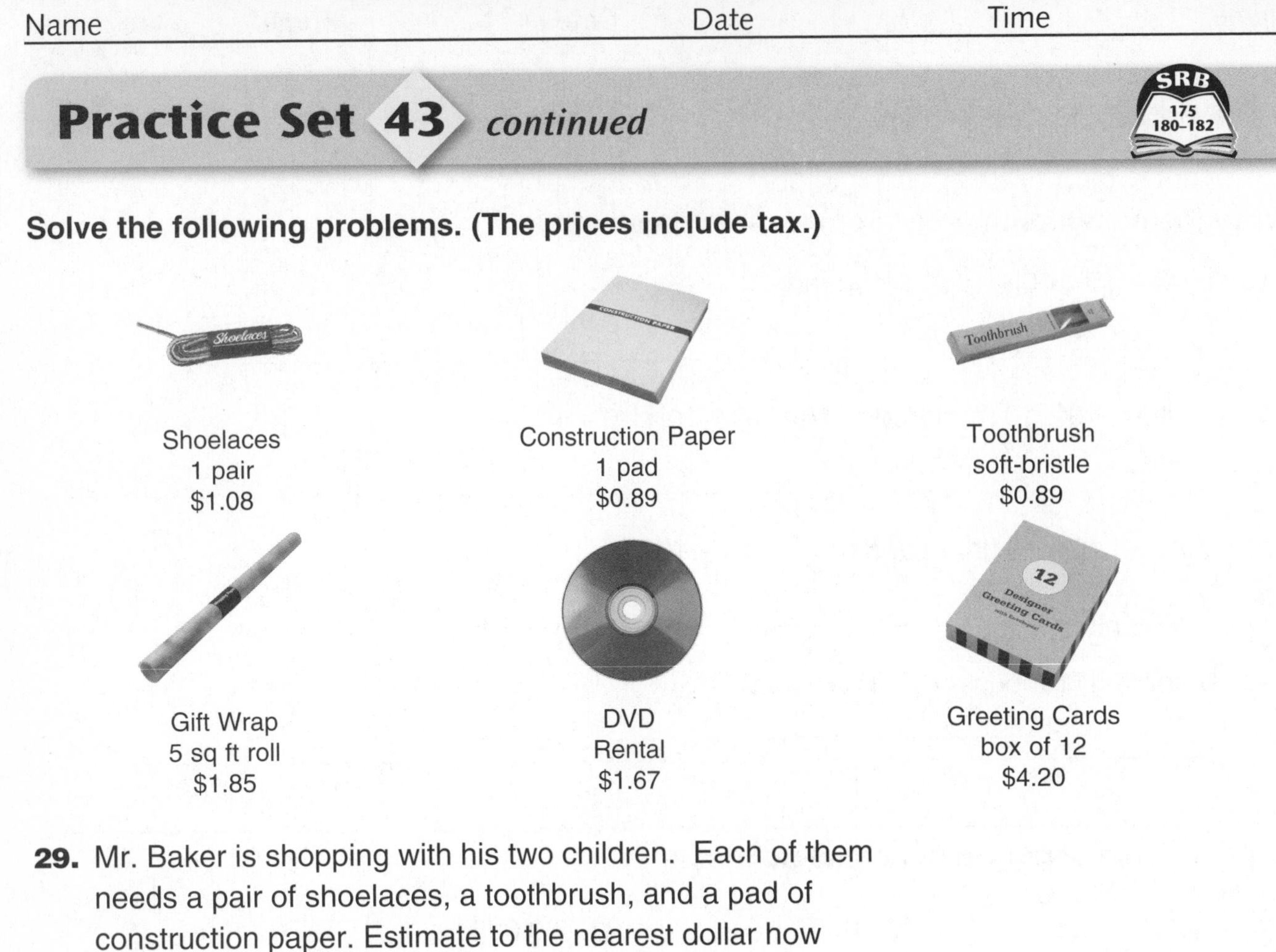

29. Mr. Baker is shopping with his two children. Each of them needs a pair of shoelaces, a toothbrush, and a pad of construction paper. Estimate to the nearest dollar how much money he will need. Show how you estimated.

____

____

30. Joe and Ahmed need to rent 2 DVDs from their city's historical society. How much money do they need?

____

31. How much money does each of the greeting cards in the box of 12 cost?

____

32. **Writing/Reasoning** Brianna needs enough wrapping paper to cover 5 presents. She has 6 dollars, and she estimates that she will need 3 rolls. Does she have enough money? Explain your answer.

____

____

Use with or after Lesson 5•9.

Name Date Time

Practice Set 44

Top 10 Longest Steel Roller Coasters in the World		
Coaster	**Park**	**Length**
Steel Dragon 2000	Nagashima Spa Land, Japan	8,133 ft
Daidarasaurus	Expoland, Japan	7,677 ft
Ultimate	Lightwater Valley, England	7,442 ft
Fujiyama	Fuji-Q High Land, Japan	6,708 ft
Millennium Force	Cedar Point, United States	6,595 ft
California Screamin'	Disney's California Adventure, United States	6,072 ft
Vertigorama	Parque de la Ciudad, Argentima	5,958 ft
Desperado	Buffalo Bill's Resort and Casino, United States	5,843 ft
Steel Force	Dorney Park, United States	5,600 ft
Mamba	Worlds of Fun, United States	5,600 ft

Use the table to answer the questions.

1. How long is the longest steel roller coaster? ____________________

2. Which country has the most roller coasters that are over 6,000 feet in length? ____________________

3. For each pair, circle the country that has the longer roller coaster.

a. United States Japan

b. England Argentina

4. How much longer is Steel Dragon 2000 than Steel Force?

Number model: ____________________ **Answer:** ________________

5. Which two roller coasters are the same length?

____

____

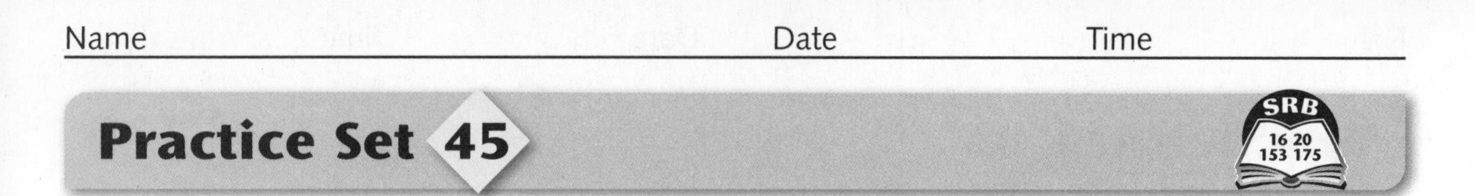

Fill in each Multiplication/Division Diagram. Then write a number model. Be sure to include a unit with each answer.

1. Caroline has 72 inches of ribbon to wrap her gifts. She has 6 gifts to wrap. How many inches of ribbon can she use for each gift?

gifts	inches of ribbon per gift	inches of ribbon in all

Number model: ______________ **Answer:** ______________

2. Kano's college is 1,200 miles away. He drives 300 miles each day. How long will it take him to get to college?

days	miles per day	miles in all

Number model: ______________ **Answer:** ______________

3. Raul organizes his CD collection in trays that hold 30 CDs. He has 270 CDs. How many trays will he need?

trays	CDs per tray	CDs in all

Number model: ______________ **Answer:** ______________

4. Hannah and two friends blow up balloons for the school fun fair. There are 84 balloons in the package. How many balloons does each person have to blow up?

people	balloons per person	balloons in all

Number model: ______________ **Answer:** ______________

Name Date Time

Practice Set 45 *continued*

SRB 16–22 98

FACTS PRACTICE

For each problem set below, do as many as you can in one minute. You can ask someone to time you.

Problem Set 1

5. $8 * 2 =$ ______ **6.** $6 * 6 =$ ______ **7.** $\frac{42}{6} =$ ______

8. $9 * 6 =$ ______ **9.** $5 * 7 =$ ______ **10.** $2 * 9 =$ ______

11. $6 * 9 =$ ______ **12.** $\frac{24}{8} =$ ______ **13.** $7 * 7 =$ ______

14. $3 * 5 =$ ______ **15.** $8 * 7 =$ ______ **16.** $5 * 5 =$ ______

17. $\frac{27}{3} =$ ______ **18.** $\frac{14}{7} =$ ______ **19.** $9 * 2 =$ ______

Problem Set 2

20. $6 * 8 =$ ______ **21.** $7 * 3 =$ ______ **22.** $\frac{54}{9} =$ ______

23. $9 * 6 =$ ______ **24.** $5 * 8 =$ ______ **25.** $7 * 9 =$ ______

26. $6 * 4 =$ ______ **27.** $\frac{30}{5} =$ ______ **28.** $4 * 4 =$ ______

29. $9 * 5 =$ ______ **30.** $7 * 4 =$ ______ **31.** $4 * 5 =$ ______

32. $\frac{27}{9} =$ ______ **33.** $\frac{16}{4} =$ ______ **34.** $9 * 9 =$ ______

Answer the questions about $\triangle ABC$.

35. What type of triangle is $\triangle ABC$?

____

36. Besides the 3 sides, what else is equal in $\triangle ABC$?

____

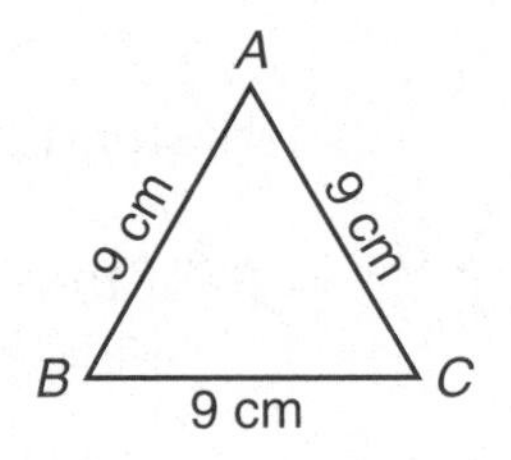

37. What is the perimeter of $\triangle ABC$?

____

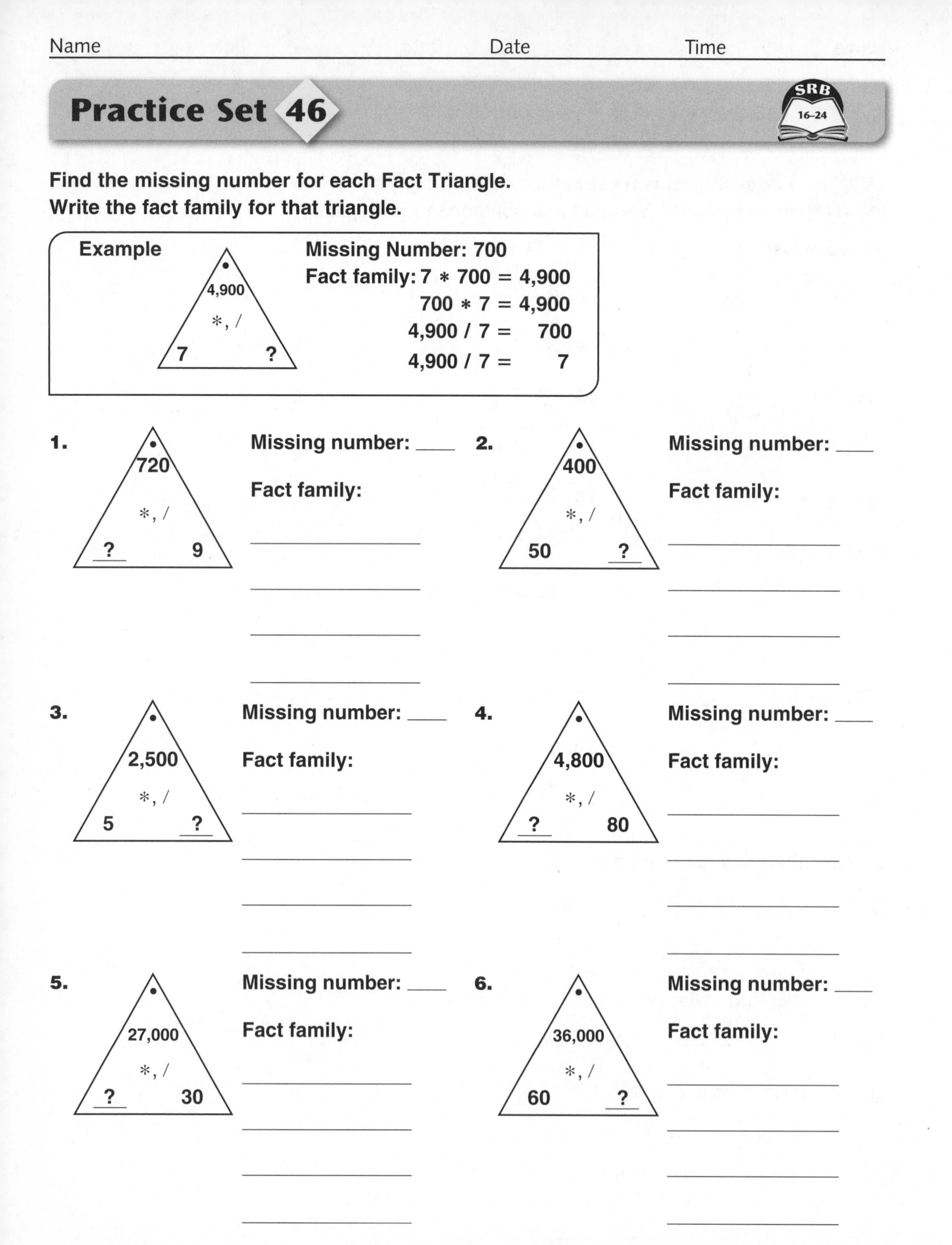

Name Date Time

Practice Set 46

SRB 16–24

Find the missing number for each Fact Triangle.
Write the fact family for that triangle.

Example

Triangle: 4,900 (top); 7, ? (bottom); $*$, /

Missing Number: 700

Fact family: $7 * 700 = 4{,}900$

$700 * 7 = 4{,}900$

$4{,}900 / 7 = 700$

$4{,}900 / 7 = 7$

1. Triangle: 720 (top); ?, 9 (bottom); $*$, /

Missing number: ____

Fact family:

______

______

______

______

2. Triangle: 400 (top); 50, ? (bottom); $*$, /

Missing number: ____

Fact family:

______

______

______

______

3. Triangle: 2,500 (top); 5, ? (bottom); $*$, /

Missing number: ____

Fact family:

______

______

______

______

4. Triangle: 4,800 (top); ?, 80 (bottom); $*$, /

Missing number: ____

Fact family:

______

______

______

______

5. Triangle: 27,000 (top); ?, 30 (bottom); $*$, /

Missing number: ____

Fact family:

______

______

______

______

6. Triangle: 36,000 (top); 60, ? (bottom); $*$, /

Missing number: ____

Fact family:

______

______

______

______

Name Date Time

Practice Set 46 *continued*

SRB 55 175 316

COMPUTATION PRACTICE

Solve.

7. 3,641 − 2,040 = ______

8. 21 * 9 = ______

9. 62 * 21 = ______

10. 178 * 5 = ______

11. 408 + 323 + 475 = ______

12. 205 + 335 + 182 = ______

13. 382 + 416 + 249 = ______

14. 414 + 627 + 100 = ______

Write the amounts.

15. (Q)(Q)(Q)(Q)(Q)(D)(D)(N)(N)(P)(P)(P) $ ______

16. [$1] [$1] [$1] (Q)(D)(D)(D)(D)(P)(P) $ ______

17. [$5] [$5] [$5] [$5] [$5] [$1] (Q)(N)(N) $ ______

18. [$100] [$100] [$20] [$20] [$5] [$1] [$1] $ ______

Write the missing numbers.

19. $\frac{1}{3}$, $\frac{2}{3}$, ______, $\frac{4}{3}$ or $1\frac{1}{3}$, ______, ______, $\frac{7}{3}$ or $2\frac{1}{3}$, ______

20. 0.5, 0.8, 1.1, ______, ______, ______, 2.3, ______

21. 2, ______, 6, 8, ______, ______, ______, ______

Name Date Time

Practice Set 47

SRB 22 23

COMPUTATION PRACTICE

Use the partial-quotients method to divide.

1. 4)124

2. 6)154

3. 9)355

4. 11)247

5. 3)195

6. 5)256

7. 14)129

8. 8)197

9. 12)232

10. 5)158

11. 7)164

12. 16)235

Name Date Time

Practice Set 47 *continued*

SRB 22 23 175

Write your answers below.

13. Eight people are going to share $168 equally.

a. How many $10 bills does each person get? ______

b. How many dollars are left to share? ______

c. From the money that remains, how many $1 bills does each person get? ______

d. What is the total amount that each person gets? ______

e. Write a number model for this problem. ______

COMPUTATION PRACTICE **Solve.**

14. 352 − 247

15. 118 ∗ 7

16. 3,276 + 1,398

17. 768 − 89

FACTS PRACTICE **In each problem set below, do as many as you can in one minute. You can ask someone to time you.**

Problem Set 1

18. 12 − 6 = ______

19. 16 ÷ 4 = ______

20. 8 ∗ ______ = 40

21. $\frac{54}{9}$ = ______

22. 5 + 3 = ______

23. 11 − 8 = ______

24. ______ ∗ 9 = 36

25. 100 ÷ 10 = ______

26. 12 ∗ 9 = ______

27. 3 ∗ ______ = 27

28. 4 + 7 = ______

29. $\frac{20}{5}$ = ______

30. 15 − 8 = ______

31. 6 + 9 = ______

32. 36 ÷ ______ = 6

Problem Set 2

33. 9 + 2 = ______

34. $\frac{32}{8}$ = ______

35. 5 ∗ ______ = 25

36. 30 ÷ 5 = ______

37. 6 + ______ = 14

38. 10 − 7 = ______

39. 64 ÷ 8 = ______

40. ______ ∗ 7 = 56

41. 4 ∗ 6 = ______

42. 4 ∗ ______ = 48

43. 16 − 7 = ______

44. 45 ÷ 5 = ______

45. $\frac{16}{4}$ = ______

46. 12 − ______ = 6

47. 7 + 6 = ______

Name Date Time

Practice Set 48

SRB 162–164 180–184

Estimate by telling whether the answer will be in the tens, hundreds, or thousands.

1. 951 ÷ 8 ______________
2. 165 / 9 ______________
3. 677 * 2 ______________
4. 92 ÷ 6 ______________
5. 924 ÷ 5 ______________
6. 472 / 15 ______________
7. 67 * 12 ______________
8. 762 ÷ 31 ______________
9. 389 ÷ 7 ______________
10. 437 ÷ 3 ______________

Write the number model. Then solve.

11. Four friends shared 516 trading cards. How many trading cards did each person get?

    Number model: ______________ **Answer:** ______________

12. Alec worked 6 days and earned $354. How much did he earn per day?

    Number model: ______________ **Answer:** ______________

13. Meredith has a collection of glass animals. She stores her collection in 3 boxes. Each box holds 24 glass animals. How many glass animals are in her collection?

    Number model: ______________ **Answer:** ______________

14. To raise money, the nature club sold 92 boxes of note cards. Each box sold for $6. How much money did the club raise?

    Number model: ______________ **Answer:** ______________

Name Date Time

Practice Set 48 *continued*

SRB 7 44 22 61 179

Write your answers below.

15. How many pieces of fruit are there? ________

16. What fraction of the fruit is bananas? ________

17. What fraction of the fruit is pears and apples? ________

18. What fraction of the fruit is bananas and pears? ________

19. What fraction of the fruit is oranges? ________

Find the missing factors.

20. 8 * ________ = 24

21. ________ * 90 = 360

22. ________ * 7 = 49

23. 2 * ________ = 960

24. 60 * ________ = 3,600

25. ________ * 7 = 350

26. 8 * ________ = 640

27. ________ * 32 = 640

Write each answer as a mixed number by rewriting the remainder as a fraction.

28. $2\overline{)31}$ 15 R1 ________

29. $10\overline{)653}$ 65 R3 ________

30. $15\overline{)455}$ 30 R5 ________

Write each answer as a decimal.

31. 37 ÷ 2 = 18 R1 ________

32. 183 ÷ 12 = 15 R3 ________

33. 2,001 ÷ 4 = 500 R1 ________

Name ______________________ Date ____________ Time ____________

Practice Set 49

SRB 6 17 18 31 92 95

First estimate and then use your full-circle protractor to measure each angle.

1.

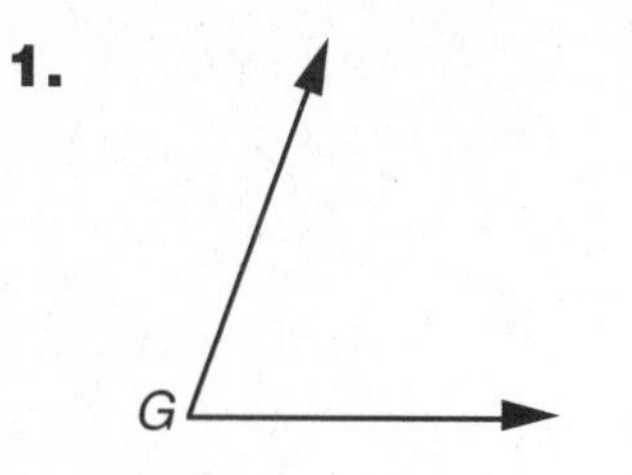

This angle is ________ (>,<) 90°.

∠*G*: ________°

2.

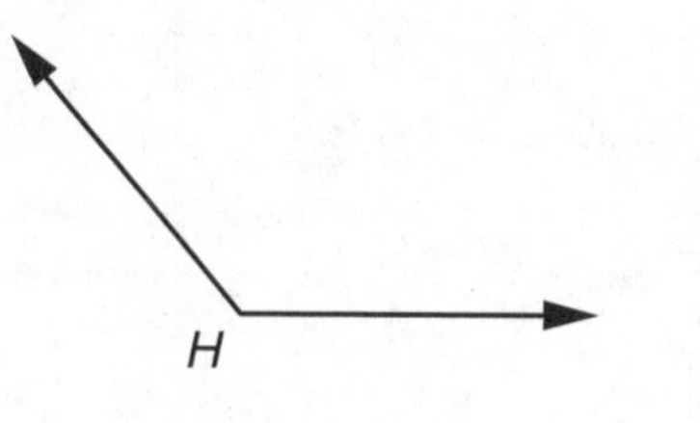

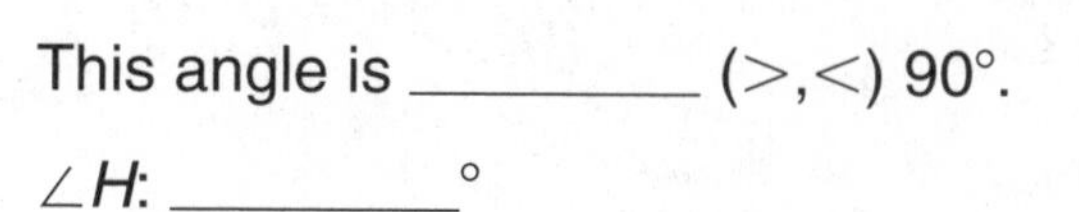

This angle is ________ (>,<) 90°.

∠*H*: ________°

3.

I

This angle is ________ (>,<) 90°.

∠*I*: ________°

4.

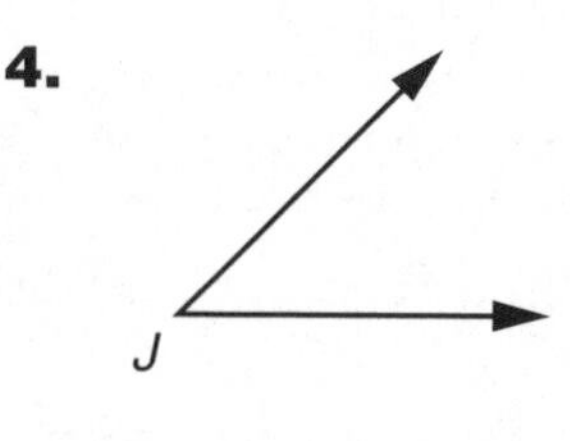

This angle is ________ (>,<) 90°.

∠*J*: ________°

Look at the picture to the right. Write *true* or *false* for each statement.

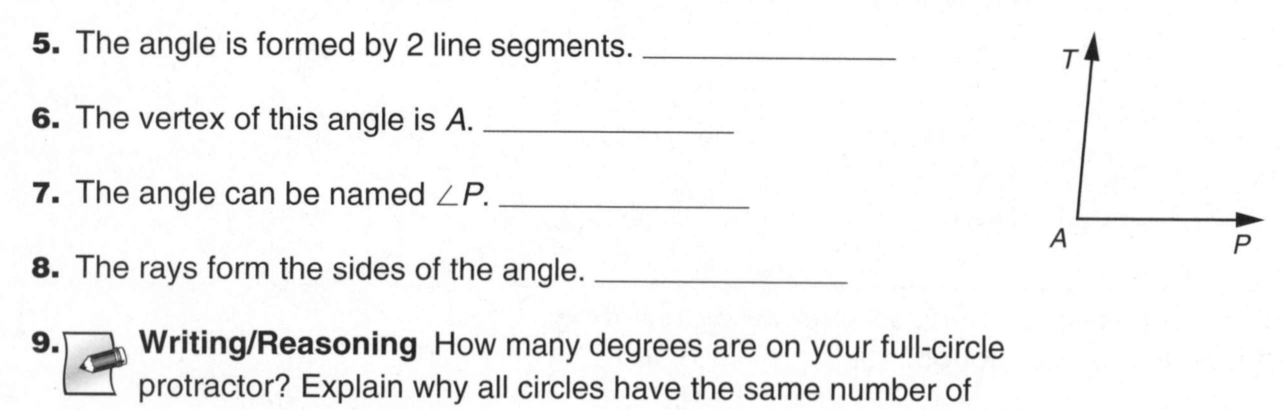

5. The angle is formed by 2 line segments. ______________

6. The vertex of this angle is *A*. ______________

7. The angle can be named ∠*P*. ______________

8. The rays form the sides of the angle. ______________

T, *A*, *P*

9. **Writing/Reasoning** How many degrees are on your full-circle protractor? Explain why all circles have the same number of degrees as your protractor.

____

____

____

Name Date Time

Practice Set 50

SRB 92 141 160 161

Write the number of degrees the minute hand moves.

1. from 1:00 to 1:15 ________
2. from 2:00 to 2:30 ________
3. from 11:00 to 11:03 ________
4. from 7:00 to 7:25 ________
5. from 10:00 to 10:25 ________
6. from 1:00 to 2:00 ________

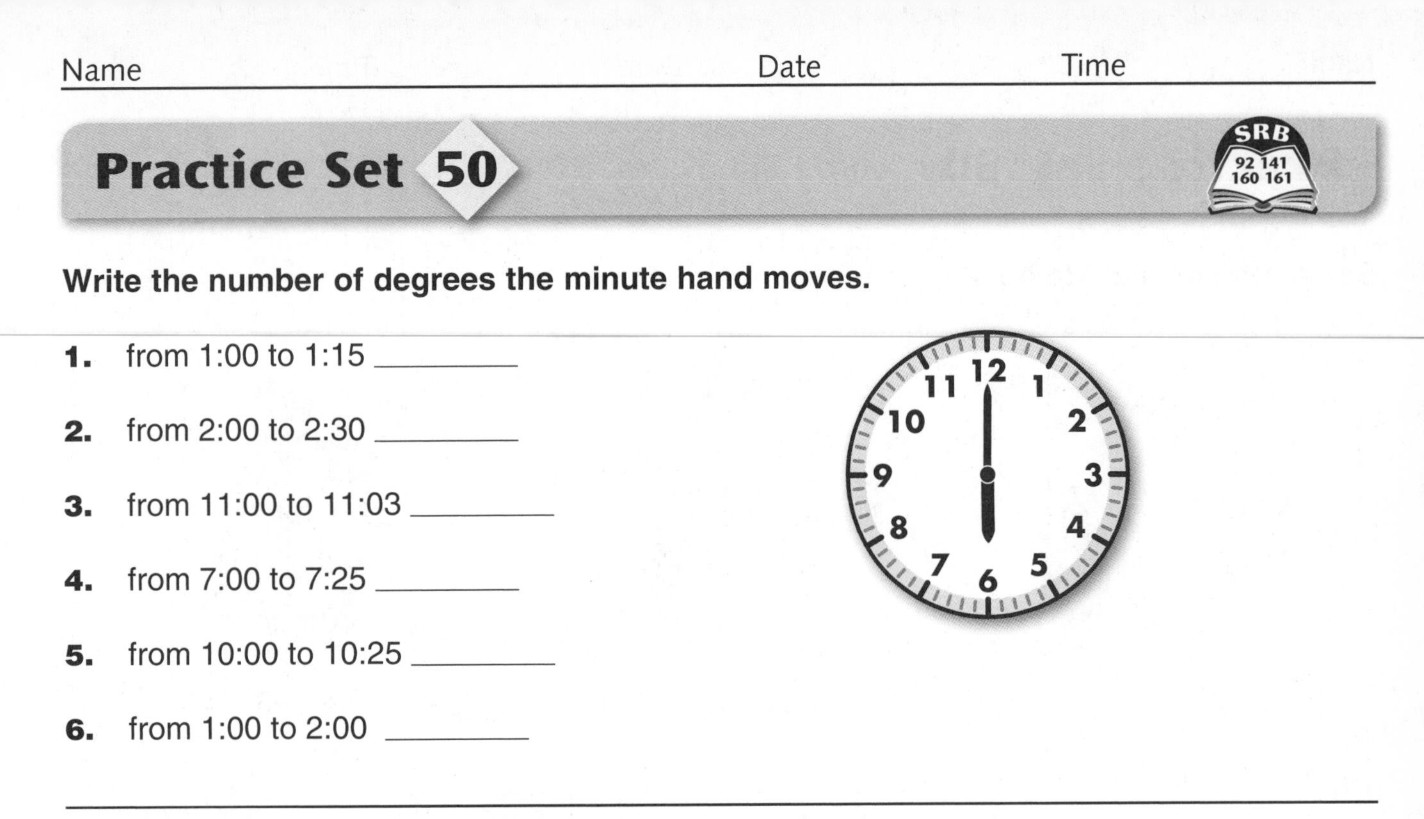

Complete the frames-and-arrows problems.

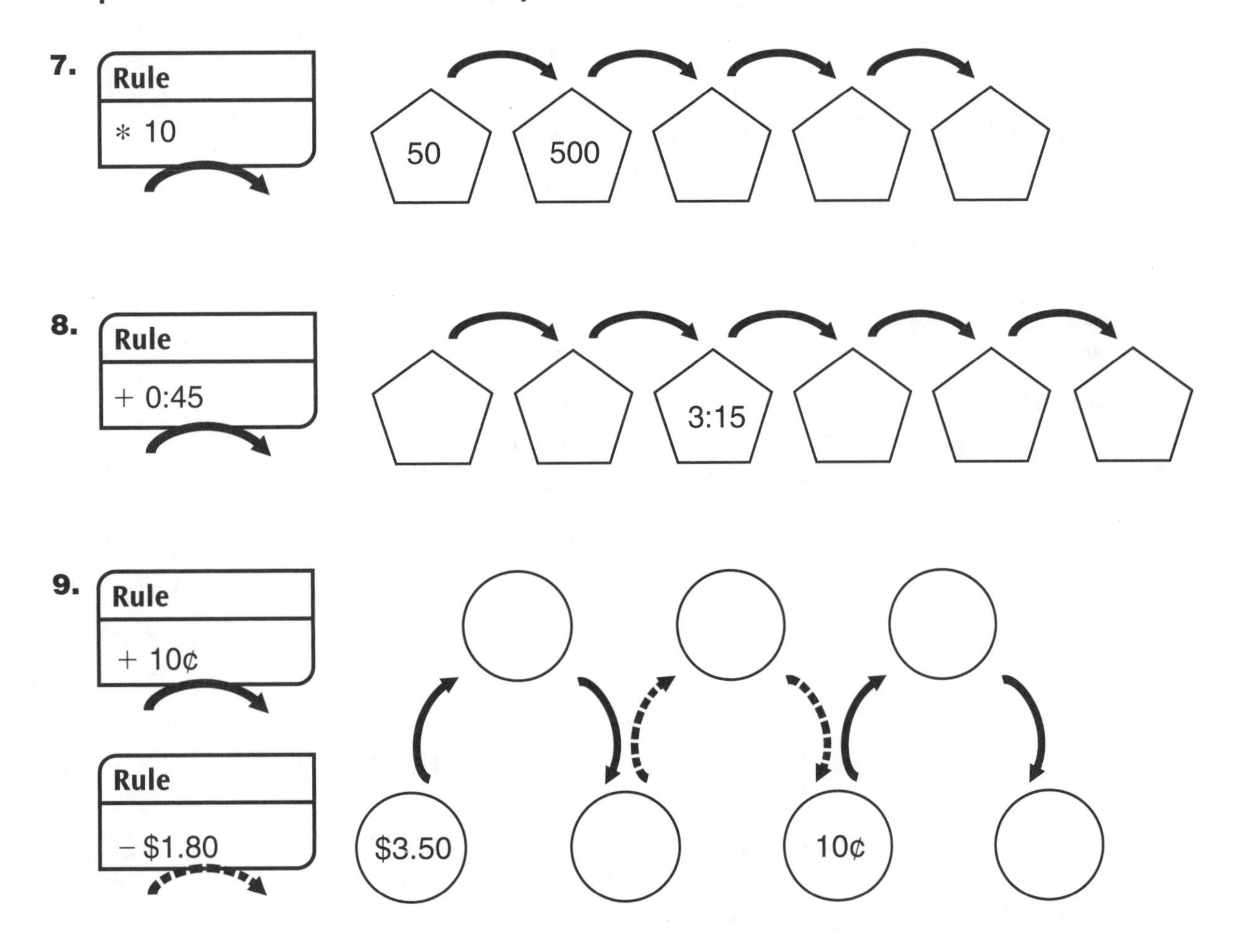

Name Date Time

Practice Set 50 *continued*

SRB 7 20 180–182

Solve. (Prices include tax.)

A $3.99

B $1.99

C $1.59

10. Ms. Jackson wants to buy enough crayons to give one crayon to each of her 29 students. She has $3.50.

a. What can she buy? ________________

b. How many crayons will she have left over? ________________

11. How many boxes of 16 crayons would it take to equal the number in the 64-crayon box? ________________

12. How much would this cost? ________________

13. **Writing/Reasoning** Estimate whether $18 is enough to buy 5 boxes of 64 crayons. Explain your reasoning.

____

____

COMPUTATION PRACTICE **Solve.**

14. 29 ∗ 3 = ________ **15.** 57 ∗ 8 = ________ **16.** 495 ∗ 6 = ________

17. 307 ∗ 4 = ________ **18.** 860 ∗ 7 = ________ **19.** 334 ∗ 11 = ________

20. Draw an array that represents the number sentence 4 ∗ 7 = 28.

Name Date Time

Practice Set 51

SRB 10–19 42–44 53

Judy brought 12 quarters to the arcade. She spent $\frac{1}{3}$ of them on video games and $\frac{1}{2}$ on basketball.

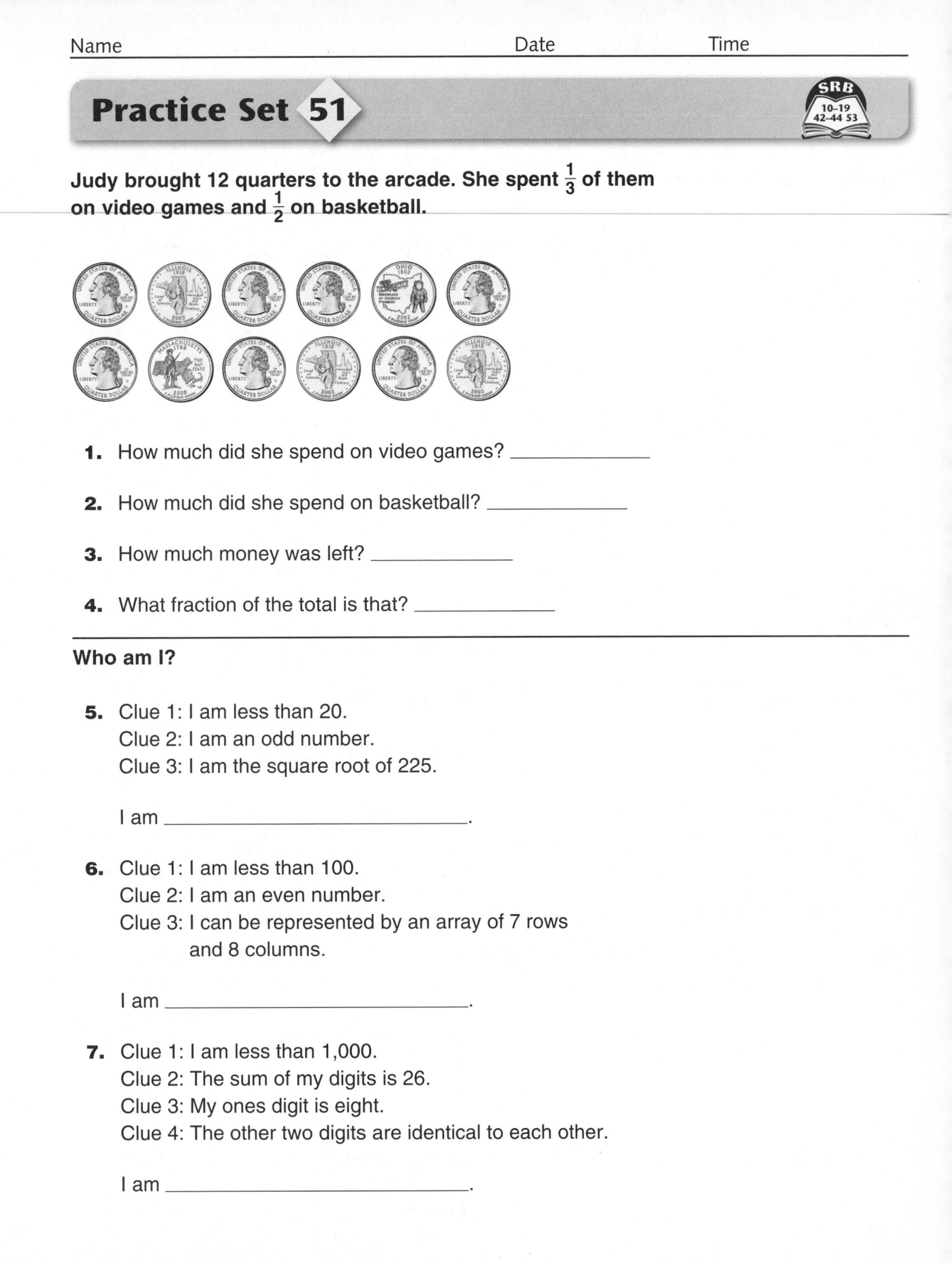

1. How much did she spend on video games? ____________
2. How much did she spend on basketball? ____________
3. How much money was left? ____________
4. What fraction of the total is that? ____________

Who am I?

5. Clue 1: I am less than 20.
   Clue 2: I am an odd number.
   Clue 3: I am the square root of 225.

   I am ______________________.

6. Clue 1: I am less than 100.
   Clue 2: I am an even number.
   Clue 3: I can be represented by an array of 7 rows and 8 columns.

   I am ______________________.

7. Clue 1: I am less than 1,000.
   Clue 2: The sum of my digits is 26.
   Clue 3: My ones digit is eight.
   Clue 4: The other two digits are identical to each other.

   I am ______________________.

Name Date Time

Practice Set 51 *continued*

SRB 153 155 162–164

Complete the "What's My Rule?" tables.

8.

Rule
out = in * 25

in	out
3	
4	
8	
	350
16	

9.

Rule
out = in / 9

in	out
81	
54	
	12
117	
	7

COMPUTATION PRACTICE **Solve.**

10. 207 − ________ = 65

11. 521 + 227 = ________

12. 190 + 448 = ________

13. 690 − 237 = ________

14. 1,416 − 948 = ________

15. 60 = 5,400 ÷ ________

16. 8,521 + 4,349

17. 1,050 * 4

18. 2,756 + 1,711

Draw a picture. Write a number model. Then solve.

19. Juan has 10 apple slices. Each friend gets three slices. How many friends get apple slices?

Number model: ________

Answer: ________

20. Jan collected 20 pennies. She shares them equally with 6 of her classmates and herself. How many pennies does each person receive?

Number model: ________

Answer: ________

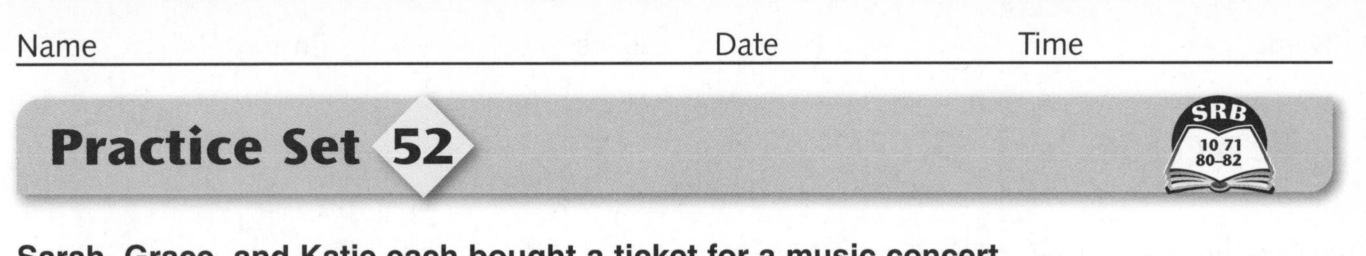

Sarah, Grace, and Katie each bought a ticket for a music concert. One ticket holder will be chosen at random to meet the band after the show. Only 500 tickets were sold. Four hundred tickets were sold to people in their town. Answer the following questions.

1. How many possible outcomes are there? ________ possible outcomes

2. Does each concert ticket have an equal chance of being drawn? ________

3. Find the probability of each event.

Event	Favorable Outcomes	Possible Outcomes	Probability
Picking Katie's ticket	1	500	$\frac{1}{500}$
Picking the ticket of one of the three girls		500	$\frac{\quad}{500}$
Picking the ticket of a person in their town		500	$\frac{400}{500}$
Picking a ticket		500	$\frac{500}{500}$

4. **Writing/Reasoning** Is it *impossible, very unlikely, likely,* or *equally likely* that one of the three girls will have her ticket drawn? Explain your answer.

____

____

Write a number model to estimate the sum.

5. 695 + 205

Number model: ____________

6. 316 + 890

Number model: ____________

7. 574 + 98

Number model: ____________

8. 412 + 1,878

Number model: ____________

Name Date Time

Practice Set 53

SRB 6 42–44 179

Write the fraction of the figure that is shaded.

1. ______

2. ______

3. ______

4. ______

5. ______

6. ______

Write > or < to make each number sentence true.

7. 9,608 ______ 9,906

8. 48,549 ______ 48,459

9. 113,012 ______ 131,102

10. 278,300 ______ 79,309

11. 3,780,576 ______ 420,777

12. 50,701,318 ______ 5,710,381

COMPUTATION PRACTICE

Divide. Write the answer as a mixed number.

13. $41 \div 6 =$ ______

14. 61 / 3 = ______

15. 87 / 4 = ______

16. $149 \div 2 =$ ______

17. 268 / 5 = ______

18. $731 \div 8 =$ ______

19. $680 \div 9 =$ ______

20. 425 / 7 = ______

Name Date Time

Practice Set 53 *continued*

SRB 48 70–75

People living in an apartment building were asked to identify the number of members in their households. The tallies in the table show the results of the survey. Use the table to help you answer the questions below.

People per Household	Number of Households
1	///
2	𝍸
3	𝍸 //
4	𝍸 /
5	//
6	//

21. How many households were interviewed?

____

22. How many people live in the building?

____

23. How many households have 4 or more people?

____

24. What is the median number of people per household in the building?

____

25. What is the approximate mean number (average) of people per household in the building to the nearest whole number?

____

26. What is the range of people per household?

____

27. What is the mode? _______________

28. **Writing/Reasoning** Explain how you found the answer to Problem 24.

____

____

____

Name Date Time

Practice Set 54

SRB 16 20 55–57

Solve.

1. A cake was cut into eight pieces.

a. Monday night $\frac{5}{8}$ of the cake was eaten. How much was left?

____

b. Marilyn ate $\frac{1}{4}$ of the cake and Billy ate $\frac{1}{8}$ of the cake. How much cake did they eat in all?

____

2. Rita draws a line segment $2\frac{1}{2}$ inches long. Then she erases $\frac{3}{8}$ inch. How long is the line segment now?

____

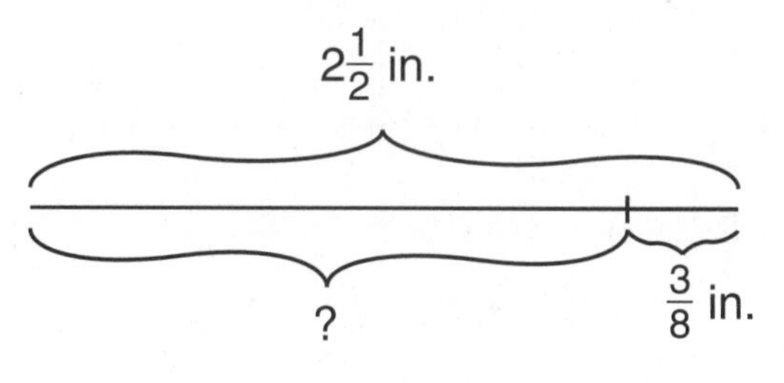

Write two multiplication and two division facts for each Fact Triangle.

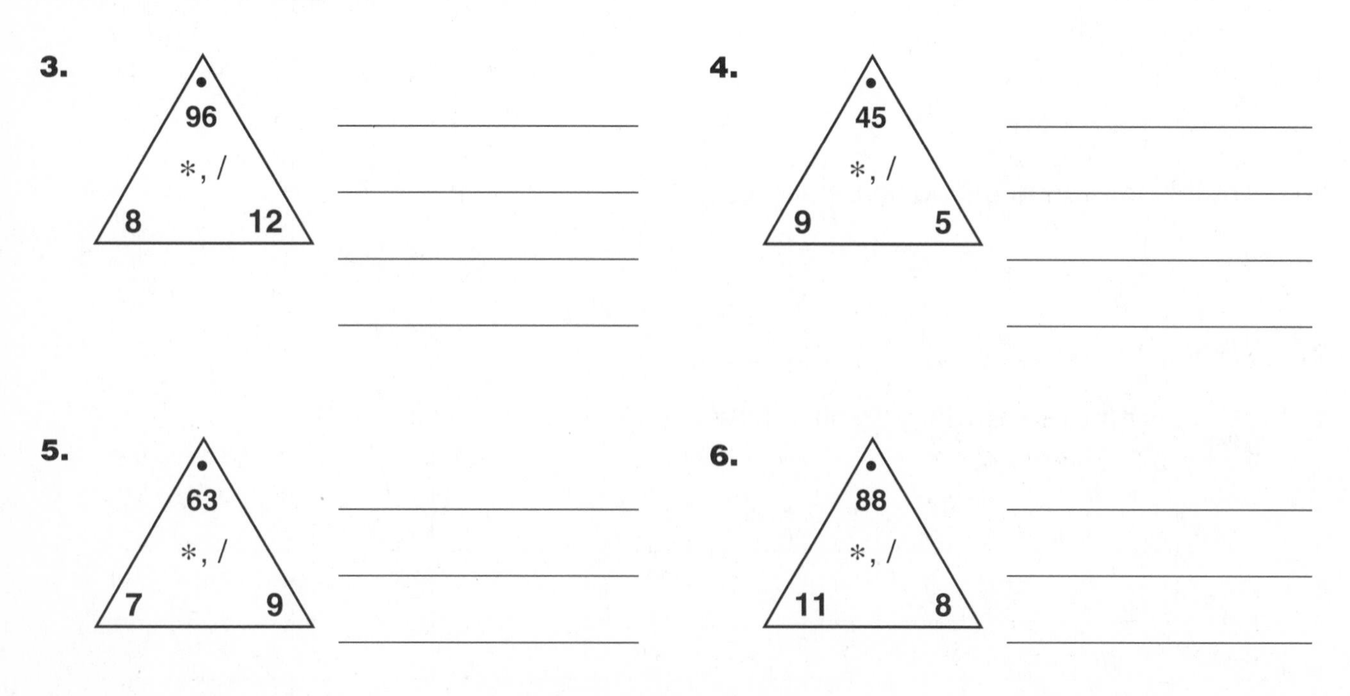

Use with or after Lesson 7•5.

Name Date Time

Practice Set 54 *continued*

SRB 42–46 175

COMPUTATION PRACTICE **Solve.**

7. $16 * 24$

8. $91 - 35$

9. $31 * 42$

10. $7\overline{)42}$

11. 486 ÷ 18

12. $16\overline{)336}$

13. $185 - 78$

14. $748 + 546$

15. $87 * 12$

16. 496 / 8

17. $79 * 57$

18. $9,017 + 4,526$

19. Daniel planted 6 rows of beets in his garden, with 24 beets in each row. How many beets did he plant in all? ______________

20. The tank of Mr. Washington's car holds about 16 gallons of gasoline. About how many gallons are in the tank when the gauge shows $\frac{1}{2}$ full? ______________

21. When the gas tank is about $\frac{1}{4}$ full, Mr. Washington stops to fill the tank. If gasoline costs $3.50 per gallon, about how much does it cost to fill the tank? ______________

22. Joshua records his weight every week. At the beginning of March, he weighed 87 pounds. His weekly weight changes in March were +2, −1, +1 and +3 pounds. What was Joshua's total weight change for March? How much did he weigh at the end of the month?

____

Name Date Time

Practice Set 55

SRB 10-16 49-51

Write the letter of the picture that represents a fraction that is equivalent to the given fraction.

A B C D

1. $\frac{1}{2}$ ______ **2.** $\frac{6}{10}$ ______ **3.** $\frac{4}{5}$ ______ **4.** $\frac{2}{6}$ ______

For each pair of fractions, write *yes* if the fractions are equivalent. Write *no* if the fractions are not equivalent.

5. $\frac{1}{6}, \frac{2}{12}$ ______

6. $\frac{7}{12}, \frac{3}{4}$ ______

7. $\frac{1}{2}, \frac{5}{10}$ ______

8. $\frac{1}{3}, \frac{2}{9}$ ______

9. $\frac{4}{5}, \frac{12}{20}$ ______

10. $\frac{40}{50}, \frac{8}{10}$ ______

11. $\frac{6}{8}, \frac{3}{4}$ ______

12. $\frac{1}{4}, \frac{4}{12}$ ______

COMPUTATION PRACTICE

Solve.

13. $1.20 * 5 = ______

14. 24 = 8 * ______

15. 9.2 + 6.3 = ______

16. 3 * ______ = $1.80

17. ______ ÷ 1,200 = 5

18. 14.0 − 7.4 = ______

19. 18 * ______ = 36

20. ______ ÷ 180 = 2

21. **Writing/Reasoning** Sara says the value of a fraction doesn't change if you add the same number to the numerator and denominator. Is Sara right? Explain your answer.

______

______

______

Name Date Time

Practice Set 55 *continued*

SRB 162–164

Complete the frames-and-arrows problems.

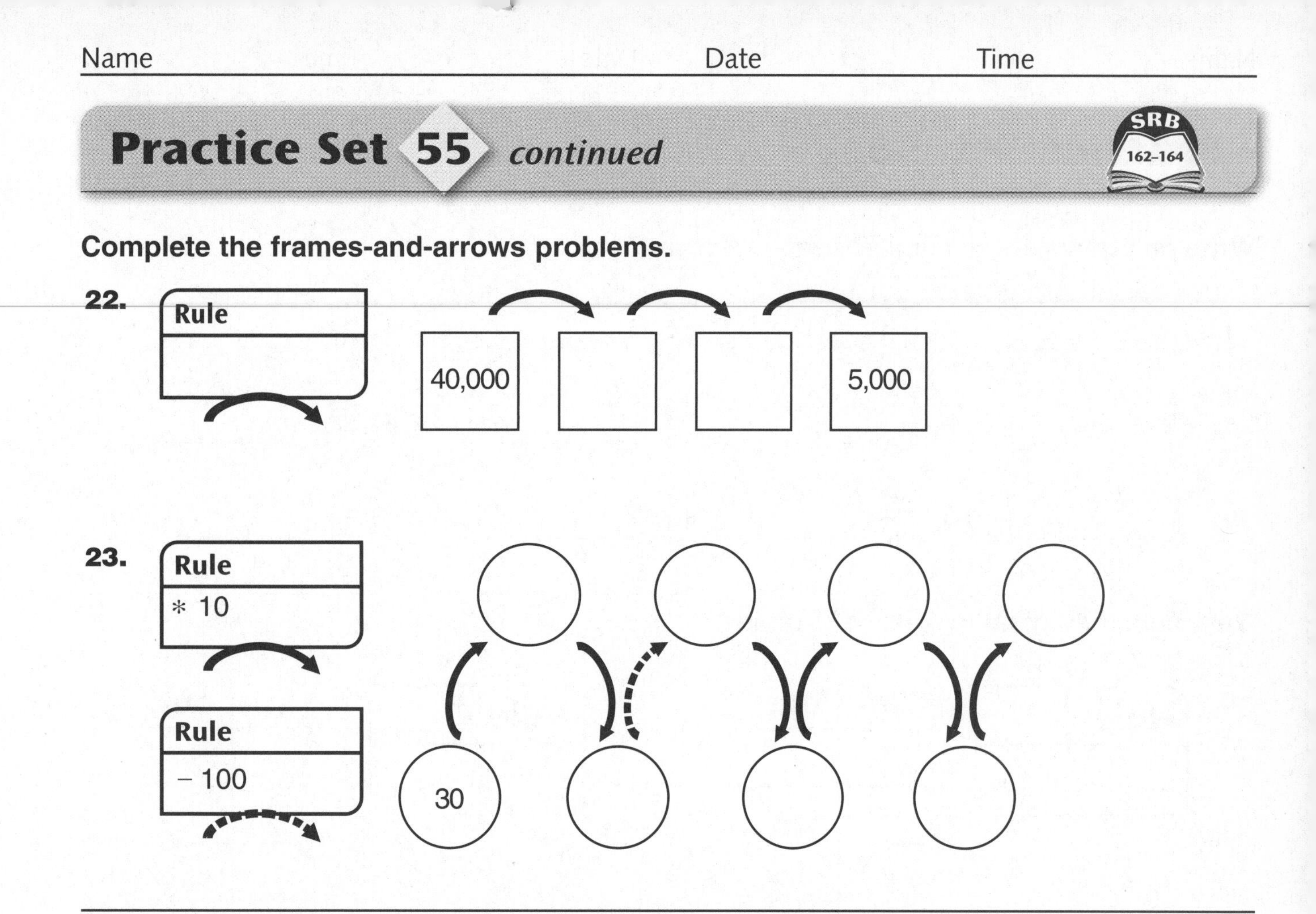

24. Plot and label the coordinate points. Then connect the points in alphabetical order. Finally, connect point *Q* to point *D*.

A (3,3) *B* (3,2)

C (5,2) *D* (6,1)

E (9,1) *F* (9,2)

G (8,2) *H* (8,6)

I (9,7) *J* (9,10)

K (8,9) *L* (5,9)

M (4,10) *N* (4,7)

O (5,6) *P* (1,2)

Q (1,1)

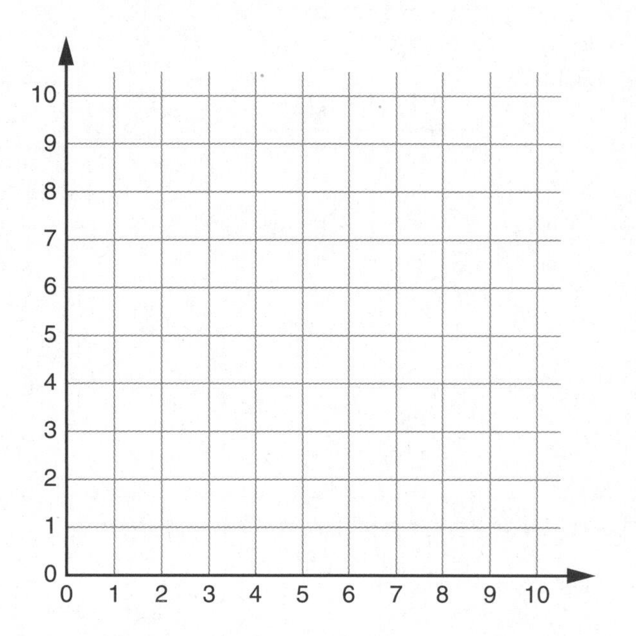

Name Date Time

Practice Set 56

SRB 55–57 61 162–164

Write an equivalent decimal for each fraction.

1. $\frac{1}{5}$ ____________

2. $\frac{74}{100}$ ____________

3. $\frac{3}{10}$ ____________

4. $\frac{98}{100}$ ____________

5. $\frac{4}{10}$ ____________

6. $\frac{4}{5}$ ____________

7. $\frac{39}{100}$ ____________

8. $\frac{19}{100}$ ____________

9. $\frac{7}{10}$ ____________

10. $\frac{55}{100}$ ____________

11. $\frac{2}{5}$ ____________

12. $\frac{81}{100}$ ____________

Complete the "What's My Rule?" tables.

13.

Rule
out = in + 15

in	out
18	
34	
	56
48	
90	

14.

Rule
out = in − 22

in	out
43	
34	
	9
18	
	77

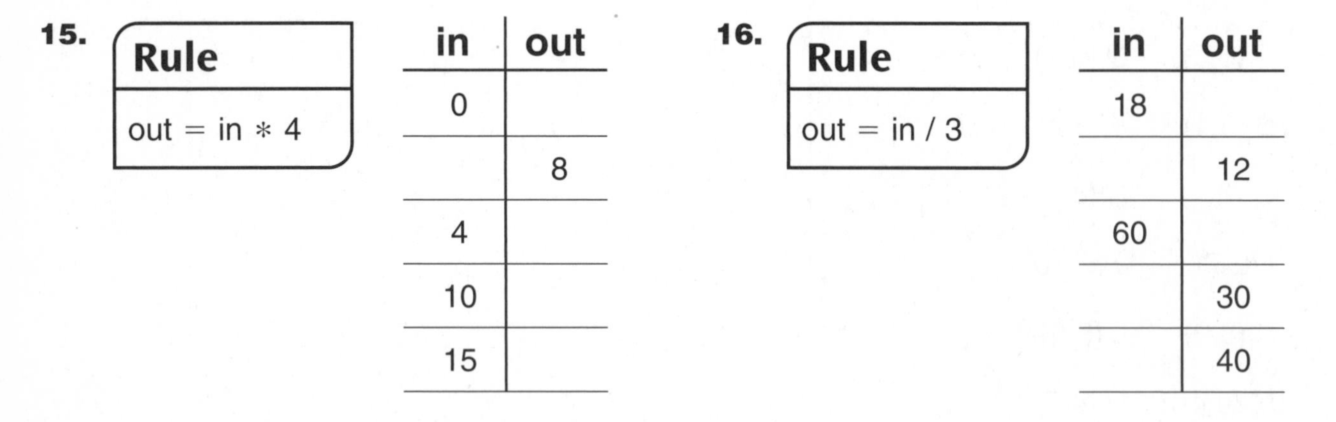

15.

Rule
out = in * 4

in	out
0	
	8
4	
10	
15	

16.

Rule
out = in / 3

in	out
18	
	12
60	
	30
	40

Solve. Use pattern blocks or pictures to help you.

17. $\frac{1}{6} + \frac{5}{6} =$ ____________

18. $\frac{2}{3} + \frac{1}{6} =$ ____________

19. $\frac{3}{4} + \frac{1}{2} =$ ____________

20. $\frac{1}{4} + \frac{1}{2} =$ ____________

Name Date Time

Practice Set 57

SRB 4 53 54

Write $<$, $>$, or $=$ to make each number sentence true.

1. $\frac{2}{5}$ ______ $\frac{7}{8}$
2. $\frac{1}{3}$ ______ $\frac{2}{3}$
3. $\frac{7}{8}$ ______ $\frac{1}{2}$
4. $\frac{5}{8}$ ______ $\frac{10}{16}$
5. $\frac{9}{10}$ ______ $\frac{2}{10}$
6. $\frac{6}{12}$ ______ $\frac{3}{6}$
7. $\frac{6}{30}$ ______ $\frac{3}{15}$
8. $\frac{9}{12}$ ______ $\frac{1}{5}$

Tell whether each fraction is less than $\frac{1}{2}$, equal to $\frac{1}{2}$, or greater than $\frac{1}{2}$.

9. $\frac{3}{6}$ ____________________
10. $\frac{10}{25}$ ____________________
11. $\frac{12}{15}$ ____________________
12. $\frac{10}{20}$ ____________________
13. $\frac{7}{10}$ ____________________
14. $\frac{1}{12}$ ____________________
15. $\frac{9}{16}$ ____________________
16. $\frac{12}{30}$ ____________________

17. Write the number that has

    4 in the tens place

    7 in the hundred-thousands place

    5 in the ones place

    0 in the thousands place

    6 in the hundreds place

    8 in the ten-thousands place

    ______ ______ ______, ______ ______ ______

Name Date Time

Practice Set 57 *continued*

SRB 10–24 154

COMPUTATION PRACTICE **Solve.**

18. $125 + 76$

19. $43 * 4$

20. $610 * 8$

21. $810 - 271$

22. $680 - 596$

23. $47 + 72$

24. $1,015 - 450$

25. $600 - 310$

26. $204 + 329$

27. $8,134 + 3,538$

28. $2,000 - 199$

29. $460 - 280$

30. It takes Sam about 35 minutes to get ready for school. If the bus comes by at 7:45 A.M., what time should Sam get up?

____

FACTS PRACTICE **Solve.**

31. 6 * ______ = 18

32. 3 * 7 = ______

33. 16 / 4 = ______

34. 20 / ______ = 4

35. 24 / 8 = ______

36. 9 * 4 = ______

37. 8 * ______ = 64

38. 81 = ______ * 9

39. 2 * 30 = ______

40. 540 ÷ ______ = 90

Name Date Time

Practice Set 58

SRB 44 59 144 184

Complete.

1. If 9 counters are $\frac{1}{2}$, then _______ counters are ONE.

2. If 4 counters are $\frac{1}{3}$, then _______ counters are ONE.

3. If 7 counters are $\frac{1}{5}$, then _______ counters are ONE.

4. If 10 counters are $\frac{2}{9}$, then _______ counters are ONE.

5. If 6 counters are $\frac{2}{3}$, then _______ counters are ONE.

6. If 15 counters are $\frac{5}{8}$, then _______ counters are ONE.

Tell if each product is in the tens, hundreds, or thousands.

7. 5 * 47 _______________

8. 22 * 95 _______________

9. 6 * 850 _______________

10. 3 * 21 _______________

Write *true* or *false* for each number sentence.

11. 82 + 7 = 99 __________

12. (5 * 2) + 6 = 60 __________

13. (27 / 9) − 15 = 18 __________

14. 81 + 5 < 100 __________

Name the ordered pair for each point plotted on the coordinate grid.

15. *A* (_______, _______)

16. *B* (_______, _______)

17. *C* (_______, _______)

18. *D* (_______, _______)

19. *E* (_______, _______)

Name Date Time

Practice Set 59

SRB 29–31 45 80–84

Use the spinner for Problems 1–5. Write *true* or *false* for each statement.

1. The chances of a paper clip landing on red are 3 out of 8. ___________
2. A paper clip is 3 times as likely to land on red as on yellow. ___________
3. A paper clip is least likely to land on blue. ___________
4. A paper clip has the same chance of landing on blue as on yellow. ___________
5. A paper clip has a $\frac{1}{6}$ chance of landing on green. ___________

red
blue
red
blue
red
green
yellow
yellow

Write the digit in the hundredths place for each decimal below.

6. 0.108 _______ 7. 13.313 _______ 8. 5.925 _______ 9. 4.078 _______

Write 2 multiplication and 2 division facts for each group of numbers.

10. 4, 9, 36

____
____
____
____

11. 6, 8, 48

____
____
____
____

12. 10, 7, 70

____
____
____
____

13. 7, 6, 42

____
____
____
____

COMPUTATION PRACTICE **Solve.**

14. 82 − 17 = ___________
15. 70 + 26 = ___________
16. 40 − 17 = ___________
17. 16 − 8 = ___________
18. 9 ∗ 9 = ___________
19. 120 − ___________ = 30

Name Date Time

Practice Set 60

SRB 131 145 158

Find the perimeter of each triangle. Convert measures of 12 inches or more to feet and inches.

1. Triangle with sides 4', 6', 8'

____

2. Triangle with sides 4' 2", 10' 1", 8' 7"

____

3. Triangle with sides 3' 6", 5' 10", 4' 8"

____

4. Triangle with sides 13' 6", 29', 35' 6"

____

If 1 centimeter on a map represents 8 kilometers, write the distance represented by each measure.

5. 2 cm _______ **6.** 3 cm _______ **7.** 10 cm _______

8. 0.5 cm _______ **9.** 2.5 cm _______ **10.** 8.5 cm _______

11. Put these numbers in order from least to greatest.

14,001 114,000 110.41 41,000

_______, _______, _______, _______

Write the next three numbers in each pattern.

12. 4, 8, 16, _______, _______, _______

13. 85, 90, 95, _______, _______, _______

14. 16, 12, 8, _______, _______, _______

15. 2, 0, −2, _______, _______, _______

Write the decimal equivalent of each fraction.

16. $\frac{50}{100}$ = _______ **17.** $\frac{77}{100}$ = _______ **18.** $\frac{48}{100}$ = _______

19. $\frac{9}{100}$ = _______ **20.** $\frac{1}{4}$ = _______ **21.** $\frac{1}{2}$ = _______

Name Date Time

Practice Set 61

SRB 129 131 145

Find the perimeter of each figure.

1.

Rectangle with sides labeled 12', 10', 10', 12'

______

2.

Triangle with sides labeled 5", 5", height 4", base 6"

______

The scale for a map is 1 inch: 20 miles. Find the distance represented by each measurement.

3. 2 inches ______

4. $\frac{1}{2}$ inch ______

5. 5 inches ______

6. $10\frac{1}{2}$ inches ______

7. $8\frac{1}{4}$ inch ______

8. 16 inches ______

COMPUTATION PRACTICE

Solve.

9. $332 - 140$

10. $38 * 8$

11. $1{,}294 + 5{,}729$

12. $600 * 50$

13. $702 + 125 + 311$

14. $39 + 67$

15. $44 + 35$

16. $92 - 48$

17. $50 - 16$

18. $87 - 36$

19. $73 - 58$

20. $509 - 376$

Name Date Time

Practice Set 62

SRB 73 133–136

Find the area of each polygon in square units.

1.

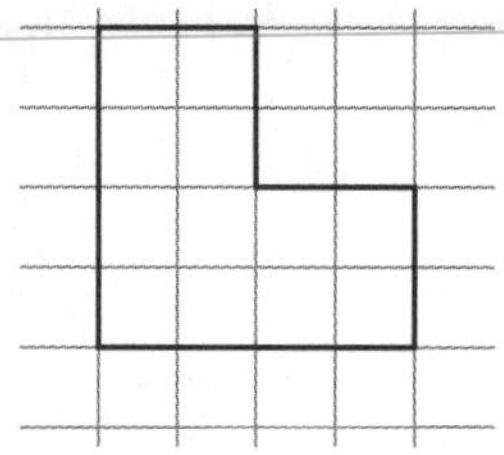

Area: ______ square units

2.

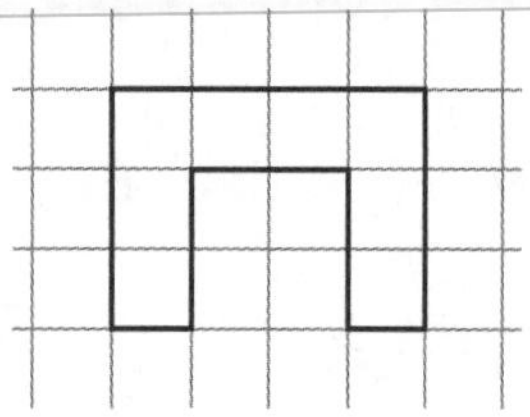

Area: ______ square units

3.

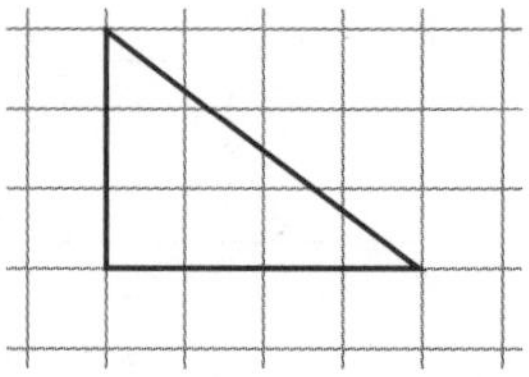

Area: ______ square units

4.

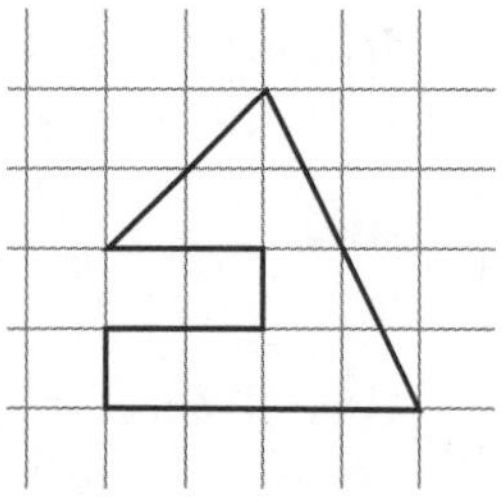

Area: ______ square units

Use the following list of numbers to answer the questions.

18, 6, 7, 9, 11, 4, 14, 8, 11, 3, 6, 11

5. Which number is the smallest? ______

6. Which number is the largest? ______

7. What is the difference between the smallest and largest numbers? ______

8. Which number appears most often? ______

9. What is the mean (average) of the numbers? ______

10. What is the median of the numbers? ______

Name Date Time

Practice Set 62 *continued*

SRB 16–24

FACTS PRACTICE

For each problem set below, do as many problems as you can in one minute. You can ask someone to time you.

Problem Set 1

11. 9 * 6 = ________

12. 2 * 8 = ________

13. 4 * 5 = ________

14. 8 * 4 = ________

15. 3 * 8 = ________

16. 8 * 7 = ________

17. 49 / 7 = ________

18. 7 * 5 = ________

19. 8 / 4 = ________

20. 6 * 2 = ________

21. 45 / 9 = ________

22. 32 / 8 = ________

23. 2 * 5 = ________

24. 9 * 8 = ________

25. 6 * 3 = ________

Problem Set 2

26. 4 * 8 = ________

27. 24 / 4 = ________

28. 5 * 9 = ________

29. 2 * 7 = ________

30. 3 * 4 = ________

31. 27 / 3 = ________

32. 6 * 8 = ________

33. 6 * 6 = ________

34. 2 * 6 = ________

35. 3 * 6 = ________

36. 7 * 3 = ________

37. 9 * 9 = ________

38. 63 / 9 = ________

39. 72 / 8 = ________

40. 81 / 9 = ________

Problem Set 3

41. 18 / 6 = ________

42. 7 * 7 = ________

43. 54 / 9 = ________

44. 64 / 8 = ________

45. 20 / 5 = ________

46. 9 * 5 = ________

47. 7 * 4 = ________

48. 28 / 7 = ________

49. 48 / 8 = ________

50. 32 / 4 = ________

51. 9 * 7 = ________

52. 8 * 4 = ________

53. 9 * 5 = ________

54. 72 / 9 = ________

55. 48 / 6 = ________

Name Date Time

Practice Set 63

SRB 155 157 160–164

Patrick's garden has vegetables, herbs, and flowers.

1. What is the total length of the garden? ____________

   The total width of the garden? ____________

2. What is the area of the vegetable section of the garden? ____________

	16'	6'
8'	vegetables	herbs
6'	flowers	

3. What is the area of the herb section of the garden? ____________

4. What is the area of the flower section of the garden? ____________

5. **Writing/Reasoning** Explain how you found the answer to Problem 4.

____

____

Rewrite the number sentences with parentheses to make them correct.

6. 7 * 9 – 4 = 59

   ____

7. 19 = 7 + 4 * 3

   ____

8. 31 – 14 – 5 = 12

   ____

9. 55 – 12 + 9 = 34

   ____

10. 4 * 9 + 4 * 12 = 84

    ____

11. 44 = 4 * 7 + 4

    ____

12. 6 * 9 – 3 = 36

    ____

13. 66 = 2 + 4 * 7 + 4

    ____

Name Date Time

Practice Set 63 *continued*

SRB 155 157 160–164

Write the next three numbers in each pattern.

14. 36, 33, 30, _____, _____, _____

15. 10, 25, 40, _____, _____, _____

16. 48, 42, 36, _____, _____, _____

17. 140, 125, 110, _____, _____, _____

18. Order these numbers from greatest to least.

3,200 32,000 2,300 23,000

_____, _____, _____, _____

COMPUTATION PRACTICE **Solve.**

19. _____ / 70 = 70

20. 6 * 30 = _____

21. 4,500 / _____ = 5

22. 9 * 90 = _____

23. 80 / 8 = _____

24. 30 * 80 = _____

25. _____ / 1,000 = 8

26. 1,400 / 700 = _____

27. 28 * _____ = 560

28. 800 = 8 * _____

29. 4 * 400 = _____

30. 30 * _____ = 1,500

Write a fraction to represent each picture.

31.

_____

32.

_____

33.

_____

34.

_____

Write as dollars and cents.

35. 18 dimes _____

36. 13 quarters _____

37. 35 nickels _____

38. 20 quarters and 6 dimes _____

39. Add the four amounts together. _____

Name Date Time

Practice Set 64

SRB 130 135 150 151

Find the area of each parallelogram.

1. 6 in. $3\frac{1}{2}$ in. ____________

2. 3.5m 1.2m ____________

Write the number sentences with parentheses and solve.

3. Add 25 to the difference of 115 and 63.

____

4. Subtract the sum of 18 and 32 from 158.

____

5. Add 19 to the difference of 150 and 116.

____

6. Subtract the sum of 58 and 42 from 210.

____

COMPUTATION PRACTICE **Solve.**

7. How many 25s in 300? ____________

8. How many 50s in 1,200? ____________

9. 8 * 2,000 ____________

10. 2,500 * 3 ____________

11. 1,500 * 7 ____________

12. 3,300 * 30 ____________

13. Without measuring, estimate this line segment to the nearest centimeter.

____________ ____________ cm

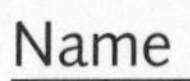
Name  Date Time

Practice Set 65

SRB 55–57 136

For Problems 1 and 2, find the area. For Problems 3 and 4, find the missing dimension.

1.

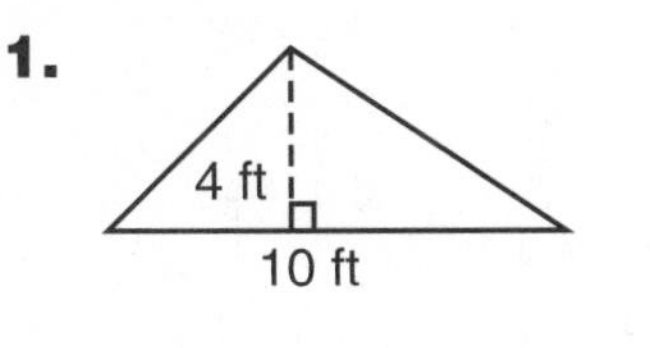

____

2.

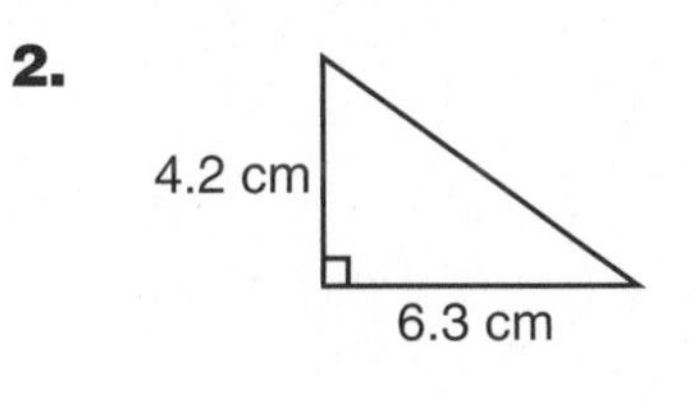

____

3. Area = 36 square meters

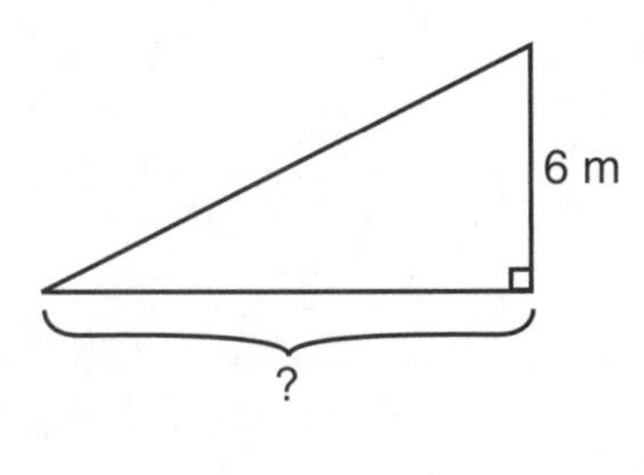

____

4. 
Area = 90 square feet

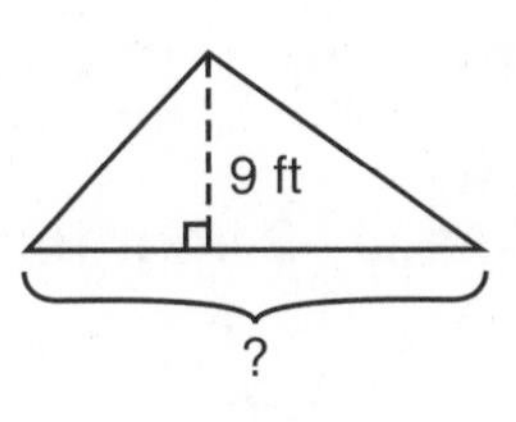

____

5. What time does the clock show? Write your answer to the nearest minute.

____

6. What time will it be in 50 minutes?

____

7. What time will it be in 128 minutes?

____

8. What time was it 2 hours and 25 minutes ago?

____

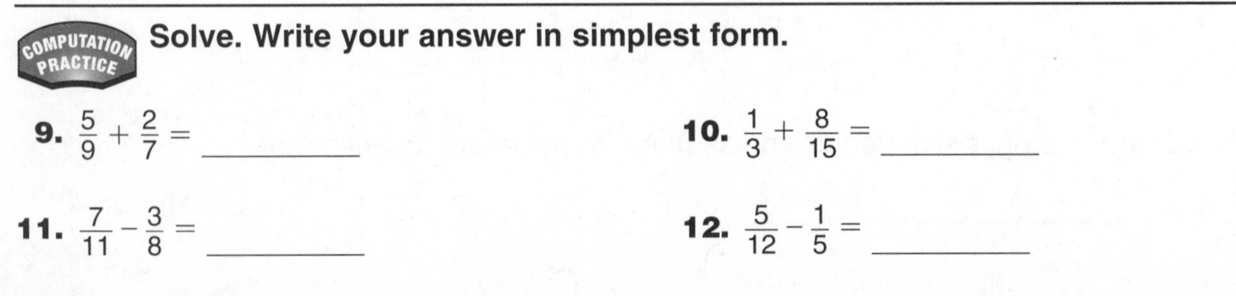

COMPUTATION PRACTICE **Solve. Write your answer in simplest form.**

9. $\frac{5}{9} + \frac{2}{7} =$ ______

10. $\frac{1}{3} + \frac{8}{15} =$ ______

11. $\frac{7}{11} - \frac{3}{8} =$ ______

12. $\frac{5}{12} - \frac{1}{5} =$ ______

Practice Set 65 *continued*

SRB 4 10–21

COMPUTATION PRACTICE

Solve.

13. 3,389 + 1,974

14. 2,974 + 189

15. 26 * 4

16. 45 * 7

17. 500 * 40

18. 1.23 + 7.91

19. 4.6 + 4.9

20. 11.40 − 6.83

21. 12 * 9

22. 58 − 22

23. 205 + 832 + 117

24. 8,362 − 4,170

25. Plot and label the coordinate points.

A (1,5)	*F* (8,7)
B (0,2)	*G* (0,0)
C (7,3)	*H* (5,1)
D (2,9)	*I* (9,6)
E (6,0)	*J* (3,8)

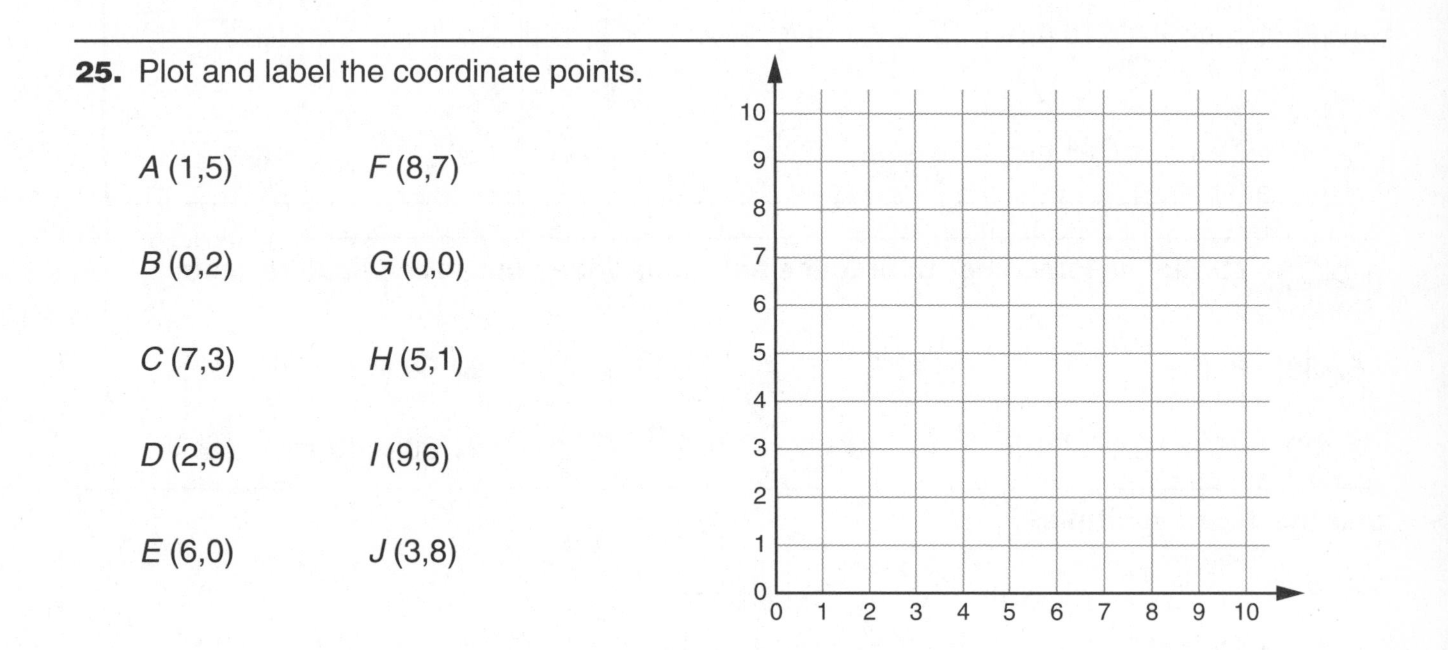

Name Date Time

Practice Set 66

Anchorage, Alaska, has a land area of 1,697 square miles. Compare this area with the area of several cities in Texas by filling in the table below. Round quotients in the last column to the nearest tenth.

Cities of Texas					
	City	Area	Area (rounded to nearest 100)	Estimate the number of times it would fit in the area of Anchorage	Divide the rounded areas. (Anchorage area ÷ city area)
1.	Houston	579 mi^2	_____ mi^2		1,697 ÷ _____ = _____ times
2.	Dallas	343 mi^2	_____ mi^2		1,697 ÷ _____ = _____ times
3.	Fort Worth	293 mi^2	_____ mi^2		1,697 ÷ _____ = _____ times
4.	Lubbock	115 mi^2	_____ mi^2		1,697 ÷ _____ = _____ times
5.	Waco	84 mi^2	_____ mi^2		1,697 ÷ _____ = _____ times

COMPUTATION PRACTICE

Multiply. Remember to practice and memorize your multiplication facts.

6. 30 * 8 = _____

7. 5 * 50 = _____

8. 60 * 3 = _____

9. 30 * 40= _____

10. 10 * 7 = _____

11. 80 * 40 = _____

List the next 5 multiples.

12. 3, _____, _____, _____, _____, _____

13. 10, _____, _____, _____, _____, _____

14. 22, _____, _____, _____, _____, _____

15. 60, _____, _____, _____, _____, _____

Name ________ Date ________ Time ________

Practice Set 67

SRB 38 49–54

Rename each decimal as a fraction with a denominator of 100 and as a percent.

1. 0.23 ________ ________ %

2. 0.52 ________ ________ %

3. 0.07 ________ ________ %

4. 0.10 ________ ________ %

Rename each percent as a fraction with a denominator of 100 and as a decimal.

5. 80% ________ ________

6. 15% ________ ________

7. 1% ________ ________

8. 24% ________ ________

Write the value of the shaded part as a decimal, a fraction, and a percent.

9.

________, ________, ________%

10.

________, ________, ________%

Write two equivalent fractions for the following numbers.

11. $\frac{1}{6}$ ________

12. $\frac{3}{7}$ ________

13. $\frac{3}{9}$ ________

14. $\frac{6}{8}$ ________

15. $\frac{11}{14}$ ________

16. $\frac{3}{5}$ ________

17. $\frac{4}{4}$ ________

18. $\frac{12}{20}$ ________

Name Date Time

Practice Set 67 *continued*

SRB 12–14

For each problem set below, do as many problems as you can in one minute. You can ask someone to time you.

Problem Set 1

19. 13 – 4 = ______

20. 13 – 8 = ______

21. 12 – 7 = ______

22. 14 – 5 = ______

23. 17 – 9 = ______

24. 12 – 8 = ______

25. 12 – 4 = ______

26. 14 – 8 = ______

27. 12 – 3 = ______

28. 16 – 9 = ______

29. 15 – 8 = ______

30. 15 – 6 = ______

31. 10 – 3 = ______

32. 11 – 8 = ______

33. 9 – 5 = ______

Problem Set 2

34. 4 – 0 = ______

35. 14 – 7 = ______

36. 9 – 9 = ______

37. 18 – 9 = ______

38. 10 – 1 = ______

39. 15 – 7 = ______

40. 13 – 7 = ______

41. 13 – 5 = ______

42. 10 – 6 = ______

43. 19 – 8 = ______

44. 12 – 5 = ______

45. 13 – 9 = ______

46. 13 – 6 = ______

47. 17 – 8 = ______

48. 11 – 7 = ______

Problem Set 3

49. 12 – 9 = ______

50. 11 – 6 = ______

51. 10 – 7 = ______

52. 16 – 8 = ______

53. 11 – 4 = ______

54. 11 – 5 = ______

55. 10 – 2 = ______

56. 15 – 9 = ______

57. 12 – 6 = ______

58. 11 – 3 = ______

59. 14 – 9 = ______

60. 16 – 6 = ______

61. 16 – 7 = ______

62. 11 – 2 = ______

63. 10 – 4 = ______

Name Date Time

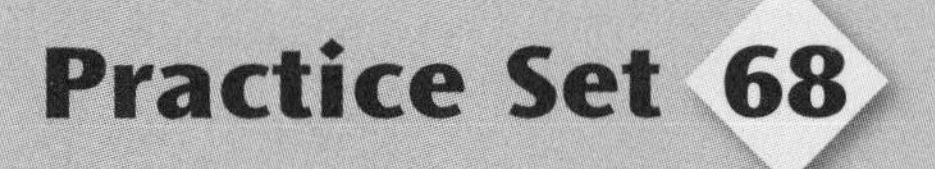

Practice Set 68

SRB 49–54 162 163

Match each decimal or fraction with the equivalent percent. Then write the letter that identifies the percent.

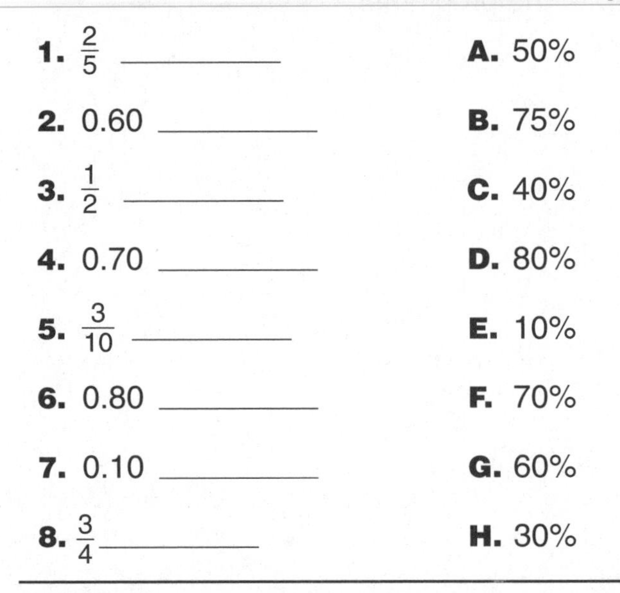

1. $\frac{2}{5}$ ________ **A.** 50%
2. 0.60 ________ **B.** 75%
3. $\frac{1}{2}$ ________ **C.** 40%
4. 0.70 ________ **D.** 80%
5. $\frac{3}{10}$ ________ **E.** 10%
6. 0.80 ________ **F.** 70%
7. 0.10 ________ **G.** 60%
8. $\frac{3}{4}$ ________ **H.** 30%

Complete the frames-and-arrows problems.

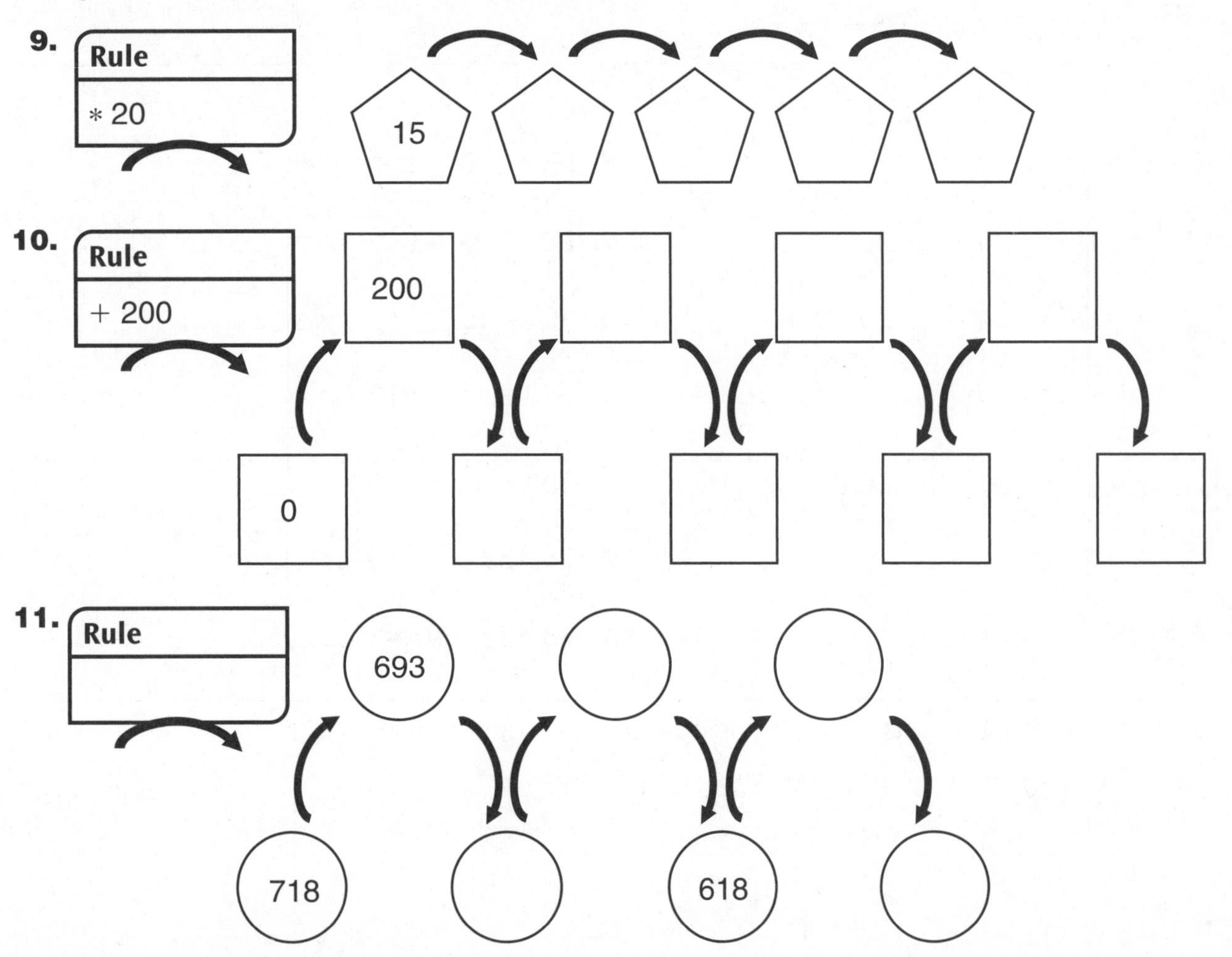

Name Date Time

Practice Set 69

SRB 4 49–54 154–155

There were 50 problems on a test.

1. Sara missed $\frac{1}{5}$ of the problems. She missed 0.20 of the problems. That's 20% of the problems.

a. How many problems did she miss? ______

b. $\frac{1}{5}$ of 50 = ______

c. 20% of 50 = ______

2. Erik missed $\frac{1}{10}$ of the problems. He missed 0.10 of the problems. That's 10% of the problems.

a. How many problems did he miss? ______

b. $\frac{1}{10}$ of 50 = ______

c. 10% of 50 = ______

Solve.

3. $40 - s = 35$ $s =$ ______

4. $55 = 18 + t$ $t =$ ______

5. $r + 9 = 51$ $r =$ ______

6. $90 - n = 43$ $n =$ ______

7. $76 + t = 206$ $t =$ ______

8. $12 = 357 - y$ $y =$ ______

9. $b - 50 = 750$ $b =$ ______

10. $543 + c = 812$ $c =$ ______

Solve these problems mentally.

11. 76,432 – 1,000 = ______

12. 76,432 – 100 = ______

13. 76,432 – 10,000 = ______

14. 76,432 – 10 = ______

---

15. Write the greatest number you can using each digit only once.

6 4 0 1 9 4 2 3 2

___ ___ ___, ___ ___ ___, ___ ___ ___

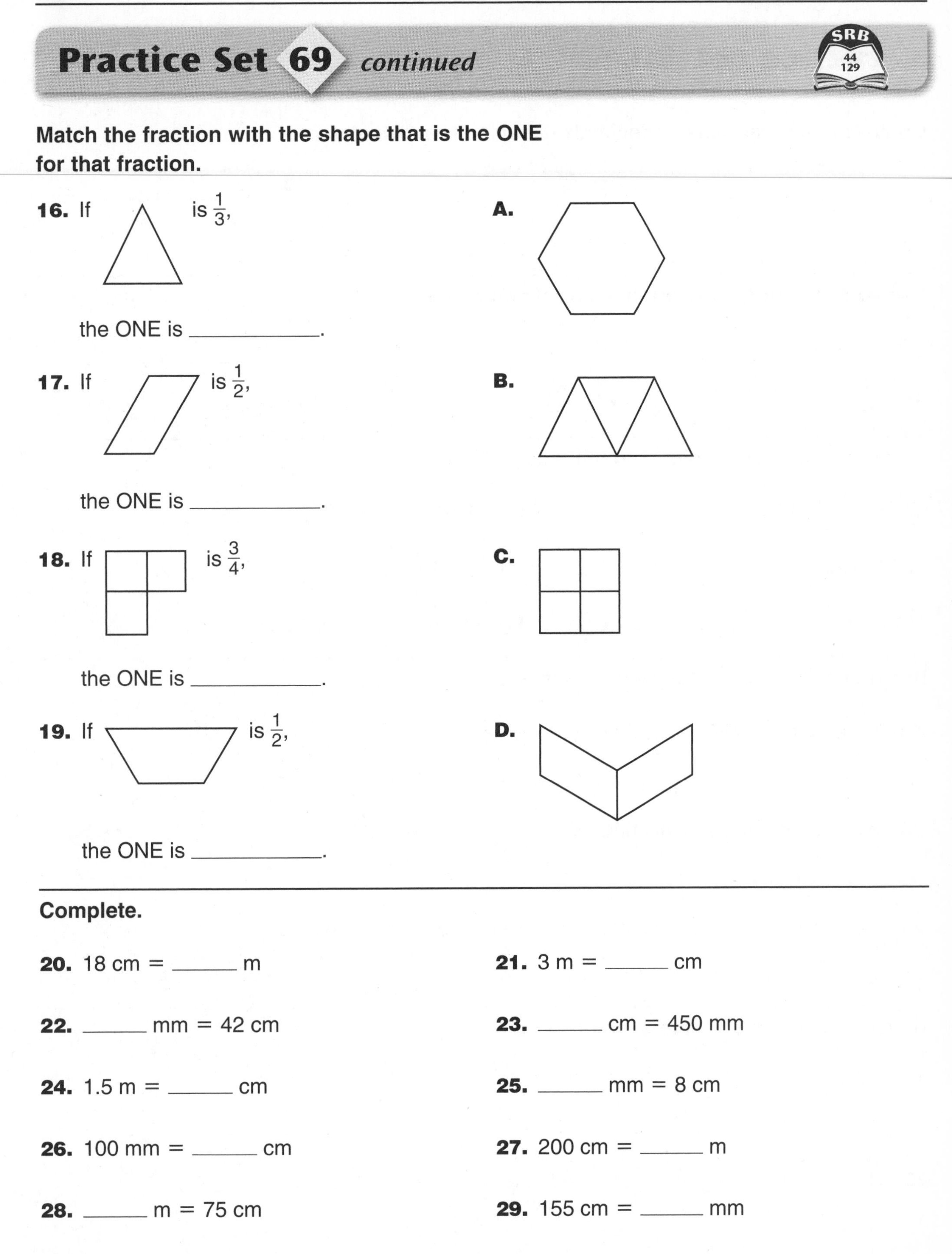

Name ______ Date ______ Time ______

Practice Set 69 *continued*

SRB 44 129

Match the fraction with the shape that is the ONE for that fraction.

16. If [triangle] is $\frac{1}{3}$,

the ONE is ______________.

17. If [parallelogram] is $\frac{1}{2}$,

the ONE is ______________.

18. If [three squares] is $\frac{3}{4}$,

the ONE is ______________.

19. If [trapezoid] is $\frac{1}{2}$,

the ONE is ______________.

A. [hexagon]

B. [three triangles]

C. [four squares]

D. [two parallelograms]

Complete.

20. 18 cm = ______ m

21. 3 m = ______ cm

22. ______ mm = 42 cm

23. ______ cm = 450 mm

24. 1.5 m = ______ cm

25. ______ mm = 8 cm

26. 100 mm = ______ cm

27. 200 cm = ______ m

28. ______ m = 75 cm

29. 155 cm = ______ mm

Name Date Time

Practice Set 70

SRB 4 61 62

Convert these fractions to decimals. Do not use a calculator.

1. $\frac{91}{100}$ ____________

2. $\frac{2}{10}$ ____________

3. $\frac{7}{20}$ ____________

4. $\frac{9}{25}$ ____________

5. $\frac{15}{50}$ ____________

6. $\frac{75}{125}$ ____________

Use a calculator to convert these fractions to percents.

7. $\frac{21}{30}$ ____________

8. $\frac{28}{70}$ ____________

9. $\frac{11}{88}$ ____________

10. $\frac{24}{80}$ ____________

11. $\frac{9}{36}$ ____________

12. $\frac{18}{48}$ ____________

Write the value of the digit 8 in the numerals below.

13. 589 ____________

14. 482,391 ____________

15. 87,402 ____________

16. 8,946,326 ____________

17. 719,538 ____________

18. 68,457 ____________

Use digits to write the following numbers.

19. seventy-four million, nine thousand, sixty-four

____

20. nineteen and sixty-eight hundredths

____

21. four hundred nine and eight hundred twenty-seven thousandths

____

Write the following numbers, using words.

22. 0.763

____

23. 18.04

____

Name Date Time

Practice Set 70 *continued*

SRB 56 96–97

Use the polygon to answer the questions.

24. Name the polygon shown. ______________

25. How many sides does it have? ______________

26. If each side were 1.5 cm long, what would the perimeter be?

____

27. **Writing/Reasoning** Explain how you found the answer to Problem 26.

____

____

Complete the "What's My Rule?" tables and rule boxes.

28. **Rule**

out = in + 38

in	out
9	
12	
15	
25	
100	

29. **Rule**

in	out
7	490
10	700
	420
11	
3	210

30. **Rule**

out = in – 108

in	out
80	
160	
	90
2,400	
	1,200

31. **Rule**

in	out
160	4
440	11
	15
800	
30	$\frac{3}{4}$

Complete.

32. 140 minutes is the same as

______________ hours and ______________ minutes.

33. 63 hours is the same as

______________ days and ______________ hours.

Name Date Time

Practice Set 71

Ms. Juarez earns $35,000 a year. Determine what percent of her money she budgets for each item.

Mrs. Juarez's Budget	
Category	**Amount**
Housing	$15,000
Clothing	$2,500
Food	$3,750
Transportation	$5,200
Savings	$3,000
Health	$3,400
Other	$2,150

1. housing __________%

2. clothing __________%

3. food __________%

4. transportation __________%

5. savings __________%

6. health __________%

7. other __________%

8. **Writing/Reasoning** Explain how you found the percents in Problems 1–7.

____

____

COMPUTATION PRACTICE **Solve.**

9. $2.20 * 5 = __________

10. 72 = 9 * __________

11. 4.2 – 2.3 = __________

12. 5 * __________ = $3.50

13. __________ ÷ 1200 = 60

14. 28.0 – 6.4 = __________

15. 60 * __________ = 4200

16. __________ ÷ 36 = 20

17. 35 = __________ ÷ 5

18. 4 * __________ = 200

Name Date Time

Practice Set 72

SRB 38 39 53–54 73

Match the expression with the number at the right. Write the letter of the number that corresponds to the expression.

1. 50% of 25,000 ________ **A.** 600

2. 10% of 6,000 ________ **B.** 5,000

3. 25% of 40,000 ________ **C.** 12,500

4. 30% of 100,000 ________ **D.** 10,000

5. 10% of 50,000 ________ **E.** 30,000

Write the fractions in order from least to greatest.

6. $\frac{1}{4}, \frac{1}{12}, \frac{1}{9}, \frac{1}{2}, \frac{1}{3}$

________, ________, ________, ________, ________

7. $\frac{4}{6}, \frac{4}{9}, \frac{4}{8}, \frac{4}{15}, \frac{4}{100}$

________, ________, ________, ________, ________

8. $\frac{15}{16}, \frac{5}{16}, \frac{3}{16}, \frac{1}{16}, \frac{7}{16}$

________, ________, ________, ________, ________

Rita checked the price of 1 pound of margarine at 5 different grocery stores. The prices she found were 85¢, 98¢, $1.09, 75¢, $1.08.

9. What is the maximum price? ________________

10. What is the minimum price? ________________

11. What is the range of prices? ________________

12. What is the median price? ________________

13. What is the mean (average) price? ________________

Name Date Time

Practice Set 72 *continued*

SRB 34–37 39

Complete.

14. $\frac{1}{4}$ of 40¢ is ________¢

15. $\frac{1}{2}$ of 50¢ is ________¢

16. $\frac{1}{3}$ of 60¢ is ________¢

17. $\frac{2}{3}$ of 30¢ is ________¢

18. $\frac{1}{3}$ of 90¢ is ________¢

19. $\frac{3}{4}$ of 80¢ is ________¢

COMPUTATION PRACTICE **Solve.**

20. 16.25 + 5.13

21. 7.4 + 5.13

22. 41.02 + 6.89

23. 5.97 + 3.6

24. 6.6 + 2.1

25. 15.43 − 3.06

26. 12.98 − 7.8

27. 59.35 − 11.63

28. 64.27 − 11.34

29. 13.89 − 2.96

Write the amounts.

30. (Q) (Q) (Q) (D) (D) (D) (D) (N) (P) (P) (P) (P) (P)

$________

31. [$1] [$1] (Q) (Q) (Q) (D) (D) (D) (D) (N) (P) (P)

$________

32. [$5] [$1] [$1] [$1] [$1] (Q) (Q) (Q) (Q) (N) (N) (N)

$________

33. [$100] [$100] [$20] [$5] [$1] [$1] [$1] (Q) (Q) (Q) (Q) (Q)

$________

Use with or after Lesson 9•7.

Name Date Time

Practice Set 73

SRB 4 27 28 153

Rewrite the product, correctly placing the decimal point.

1. 7 * 2.3 = 161 ________
2. 0.9 * 42 = 378 ________
3. 1.4 * 281 = 3,934 ________
4. 0.03 * 510 = 153 ________
5. 6.16 * 40 = 2,464 ________
6. 2,435 * 7.2 = 175,320 ________

COMPUTATION PRACTICE **Multiply.**

7. 2.6 * 96 = ________
8. 0.51 * 43 = ________
9. 5.42 * 6 = ________
10. 8.7 * 12 = ________
11. 124 * 0.9 = ________
12. 1.64 * 52 = ________

In the numeral 28,490, the 8 stands for 8,000.

13. What does the 4 stand for? ________
14. What does the 2 stand for? ________
15. What does the 9 stand for? ________
16. What does the 0 stand for? ________

The decimal point is missing in each answer. Write a number model to show how you estimate the answer. Then correctly place the decimal point in the answer.

17. 22 * 7.2 = 1584
Number model:
____

18. 6.3 * 8 = 504
Number model:
____

19. 0.33 * 60 = 1980
Number model:
____

20. 40 * 12.7 = 5080
Number model:
____

Name Date Time

Practice Set 73 *continued*

SRB 4 10–14 44 62

Tell whether each number sentence is *true* or *false*.

21. 14 + 8 = 22 ____________

22. 65 − 12 = 54 ____________

23. 18 = 39 − 14 ____________

24. 74 = 26 + 48 ____________

Write the answers.

25. What fraction of the coins is pennies? _________

26. What fraction of the coins is nickels? _________

27. What fraction of the coins is dimes? _________

28. What is the value of all the coins? _________

29. If you took away $\frac{1}{3}$ of the dimes, how much money would be left? _________

30. Use the clues to complete the place-value puzzle.

- Add 3 to the result of 71 – 68. Write the result in the hundredths place.
- Write the result of 54 / 9 in the ones place.
- Multiply 6 * 12. Subtract 65. Write the result in the tens place.
- Double the number in the ones place. Then divide by 3 and write the result in the thousandths place.
- Divide 24 by 6. Add 5 and write the result in the tenths place.

10s	1s		0.1s	0.01s	0.001s
		.			

Name Date Time

Practice Set 74

SRB 21–24 156

COMPUTATION PRACTICE

Divide.

1. 28.4 / 4 = ________

2. 1.68 / 8 = ________

3. 211.5 / 9 = ________

4. 68.4 ÷ 6 = ________

5. 11.2 / 7 = ________

6. 36.5 ÷ 5 = ________

7. Without measuring, estimate the length of this line segment to the nearest inch.

____
(Unit)

8. Make 100s.

78 + ______

53 + ______ ← 100 → 17 + ______

49 + ______

9. A square number is the product of a number multiplied by itself. For example, 25 is a square number because 5 * 5 = 25. Circle the numbers that are square numbers.

16 28 36 100 54

Estimate whether the difference is closest to 0, 1, or 2.

10. $\frac{2}{4} - \frac{1}{2}$ ________

11. $2 - \frac{3}{4}$ ________

12. $1\frac{6}{10} - \frac{3}{5}$ ________

13. $3\frac{4}{5} - 1\frac{6}{10}$ ________

Name Date Time

Practice Set 75

SRB 73–75 128 158

Tina tossed two dice 14 times and got the following sums:

6, 5, 3, 8, 2, 6, 9, 4, 11, 6, 6, 4, 8, 6

1. What is the maximum? ________

2. What is the minimum? ________

3. What is the range? ________

4. What is the median? ________

5. What is the mode? ________

6. What is the mean? ________

Fill in the missing numbers on the number lines.

7. 3 ____ ____ 48 ____ ____ 93

8. 6 ____ ____ ____ ____ 56

9. 11 ____ ____ ____ ____ ____ 53

10. 14 ____ ____ ____ 46 ____ 62

Measure the line segment to the nearest centimeter.

11. ______________________________

____ cm

12. ______________________

____ cm

Solve.

13. $88{,}574 - 9{,}695$

14. $5{,}983 + 11{,}389$

15. $96{,}312 - 45{,}160$

16. $73{,}418 - 24{,}972$

Name ______ Date ______ Time ______

Practice Set 76

SRB 80–85 106 107 109 129

Write *yes* if the image is a reflection. Write *no* if the image is not a reflection.

1. Preimage | Image

Line of Reflection

______

2. Preimage | Image

Line of Reflection

______

Complete.

3. 20 cm = ______ mm

4. 5,000 mm = ______ m

5. 20,000 mm = ______ cm

6. 2 m = ______ mm

7. 15 m = ______ cm

8. 2,000 mm = ______ cm

A threadworm is about 306 mm long.

9. What is its length in centimeters? ______ cm

10. What is its length in meters? ______ m

Suppose you spin a paper clip 180 times on the base of the spinner below.

red | blue | blue

11. How many times would you expect it to land on red?

______

12. How many times would you expect it to land on blue?

______

13. **Writing/Reasoning** Is it *likely, certain,* or *impossible* to land on green? Explain your answer.

______

______

Name Date Time

Practice Set 77

SRB 4 93 106

Write *yes* if the image is a reflection. Write *no* if the image is not **a reflection.**

1. Preimage Image

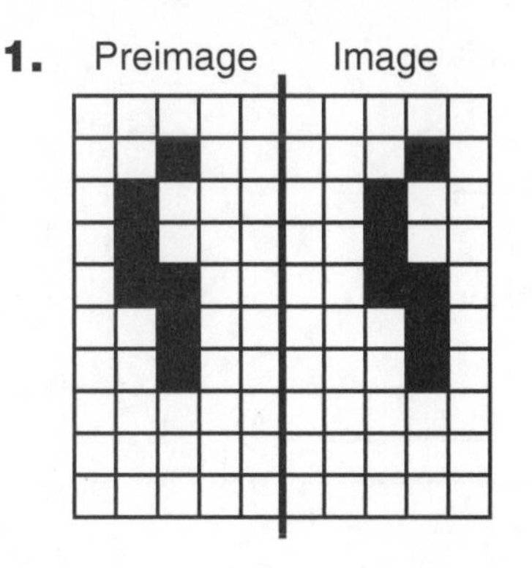

____

2. Preimage Image

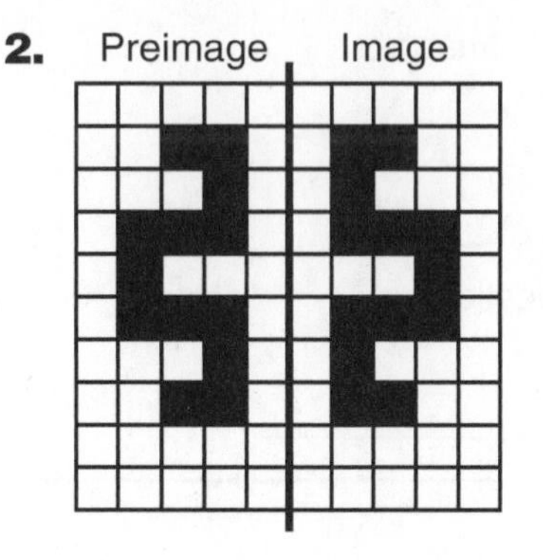

____

Write the following numbers with digits.

3. three hundred million, seventy-nine thousand

____

4. four billion, sixty-five million, seven hundred thousand

____

5. eighty-four billion, one hundred ninety-six million, forty thousand

____

Write *acute, obtuse, straight,* or *reflex* for each angle.

6.

8.

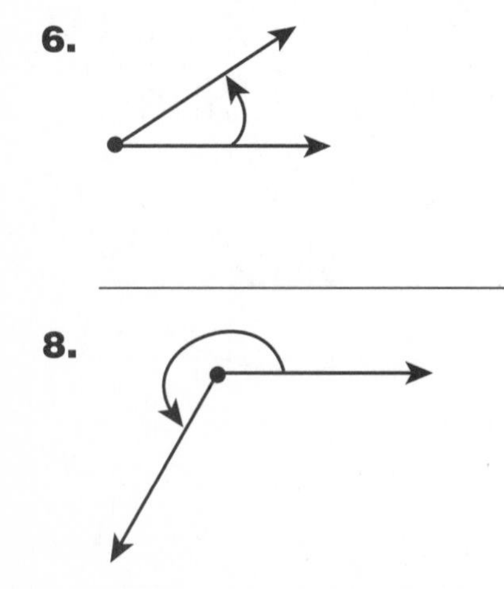

7.

9.

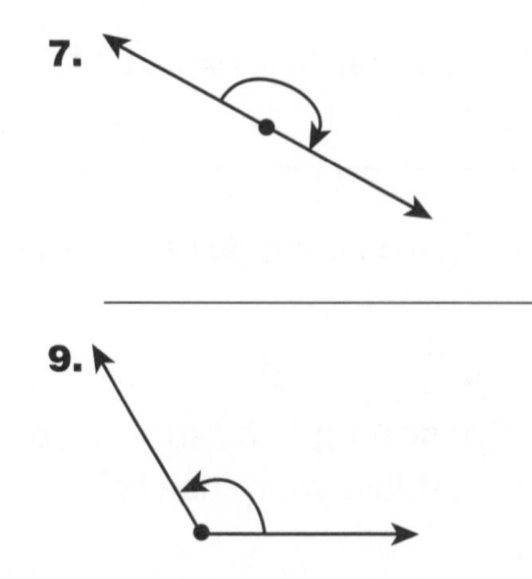

Practice Set 77 *continued*

SRB 134 149 175 176

Write number sentences for the following. Then tell whether they are *true* or *false*.

10. If 8 is subtracted from 24, the result is 16.

________________ __________

11. 6 is twice as much as 12.

________________ __________

12. 834 is more than 654.

________________ __________

13. Divide 86 by 2 and the result is 43.

________________ __________

14. If 98 is decreased by 16, the result is 84.

________________ __________

15. 27 is greater than the sum of 8 and 15.

________________ __________

Find the area in square units for each rectangle. Then write the number model.

Area = length *(l)* ∗ width *(w)*

16.

Area: ______________

Number model:

____

17.

Area: ______________

Number model:

____

Name Date Time

Practice Set 78

SRB 109 129 150–151

Write *yes* if the figure has a vertical line of symmetry. Write *no* if it does not.

1. ________

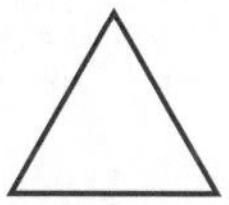

2. ________

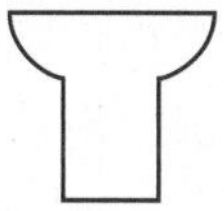

3. ________

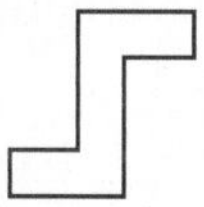

Write *yes* if the figure has a horizontal line of symmetry. Write *no* if it does not.

4. ________

5. ________

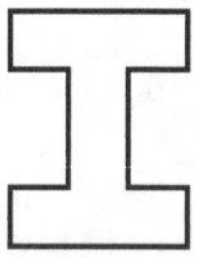

6. ________

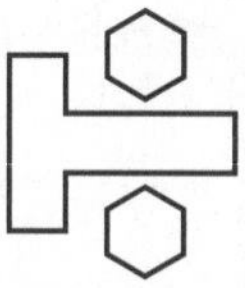

Rewrite the number sentences with parentheses to make them correct.

7. 9 * 12 – 3 = 81 ________________

8. 9 * 12 – 3 = 105 ________________

9. 15.8 = 2 * 6.5 + 2.8 ________________

10. 18.6 = 2 * 6.5 + 2.8 ________________

11. 7 * 1.1 + 4.2 * 12 = 58.1 ________________

12. 5 * 12 + 2 – 4 = 58 ________________

13. 5 * 12 + 2 – 4 = 66 ________________

14. 8,140 = 110 * 50 + 24 ________________

Complete.

15. 4 ft = ________ in.

16. 3 yd = ________ ft

17. 3 ft 5 in. = ________ in.

18. 2 yd 1 ft = ________ ft

19. 38 in. = ________ ft ________ in.

20. 9 ft = ________ yd

Name _______________ Date _______________ Time _______________

Practice Set 79

SRB 61 73 158 159

Rename the following fractions as decimals.

1. $\frac{1}{10}$ ________
2. $\frac{2}{4}$ ________
3. $\frac{6}{16}$ ________
4. $\frac{6}{10}$ ________
5. $\frac{500}{1,000}$ ________
6. $\frac{47}{100}$ ________
7. $\frac{7}{8}$ ________
8. $\frac{3}{4}$ ________
9. $\frac{9}{16}$ ________
10. $\frac{34}{100}$ ________
11. $\frac{560}{1,000}$ ________
12. $\frac{18}{18}$ ________

Tim found 5 different prices for notebooks:

35¢, $1.15, $1.29, $2.18, $1.17

13. What is the maximum price? ________
14. What is the minimum price? ________
15. What is the range of prices? ________
16. What is the median price? ________
17. What is the mean (average) price? ________

Fill in the missing numbers on the number lines.

18. 8 ____ 16 ____ 24 ____ 32 ____ ____ ____ 48

19. 0 ____ ____ 54

20. 0 ____ ____ ____ ____ ____ ____ ____ ____ ____ 200

21. 6 ____ ____ 24

Name Date Time

Practice Set 79 *continued*

SRB 42 49 50

22. How many pieces of fruit are there? __________

23. What fraction of the fruit are bananas? __________

24. What fraction of the fruit are apples? __________

25. What fraction of the fruit are lemons? __________

26. **Writing/Reasoning** Explain how you found the fraction in Problem 25.

____

____

____

COMPUTATION PRACTICE

Solve.

27. $(45 + 45) * 10 =$ ______

28. $32 - (2 * 15) =$ ______

29. $783 + 17 - 35 =$ ______

30. $(\frac{80}{4}) - 12 =$ ______

31. $50 + (\frac{75}{25}) =$ ______

32. $(30 * 6) + 22 =$ ______

33. $(\frac{90}{2} + 7) \div 2 =$ ______

34. $(16 + 4) * (3 + 2) =$ ______

35. $563 + (2 * 250) =$ ______

36. $(\frac{51}{17}) + 2 =$ ______

37. $204 \div (\frac{34}{2}) - 12=$ ______

38. $(1{,}020 \div 3) + 60 =$ ______

Name Date Time

Practice Set 80

SRB 6 139 175–178

Look at the thermometer. Answer the questions below.

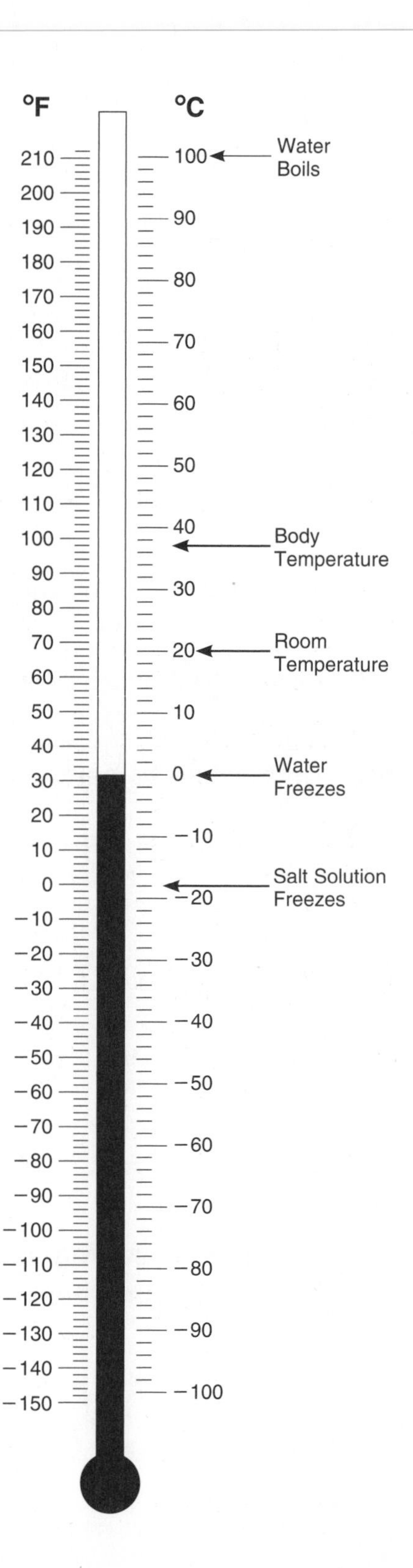

1. What is the temperature difference, in °C, between body temperature and room temperature?

____

2. What is the temperature difference, in °F, between the boiling point and freezing point for water?

____

3. What is the temperature difference, in °F, between the freezing point for water and the freezing point for a salt solution? What is the difference in °C?

_______________°F _______________°C

4. How much colder is –110°F than 7°F?

____

5. How much warmer is 42°C than –18°C?

____

6. Which is colder, –32°C or –32°F?

____

7. Which is warmer, 48°C or 108°F?

____

8. Imagine that it is 22°C outside. Which is a more likely outdoor activity: ice skating or bike riding?

____

Name Date Time

Practice Set 80 *continued*

SRB 3 44 60 158 159

Write the next three numbers in each pattern.

9. –15, –10, –5, ________, ________, ________

10. 0.04, 0.06, 0.08 ________, ________, ________

11. 0.44, 0.68, 0.92 ________, ________, ________

Julie and Pattie have 18 bananas, 16 oranges, and 20 apples. They are making bags of mixed fruit, with 4 pieces of fruit in each bag. They can put any combination of fruit in each bag.

12. How many bags can they make? ________

13. How many pieces of fruit will they have left over? ________

14. If they also had 7 pears, how many bags could they make? ________

Use the number line to help you solve Problems 15–24.

–10 –9 –8 –7 –6 –5 –4 –3 –2 –1 0 1 2 3 4 5 6 7 8 9 10

15. 5 + (–9) = ________

16. (–3) + (–4) = ________

17. (–5) + 4 = ________

18. 9 + (–3) = ________

19. (–6) + 7 = ________

20. 3 + 4 = ________

21. (–4) + 11 = ________

22. 10 + (–18) = ________

23. (–5) + (–3) = ________

24. **Writing/Reasoning** Explain how you used the number line to find the answer to Problem 19.

____

____

Name Date Time

Practice Set 81

SRB 16–24 140 180

A nickel weighs about 5 grams. A liter of soda weighs about 1 kilogram.

Match the object with a possible weight. Write the letter of the possible weight.

1. a pair of scissors ______	**A.** 200 kg
2. a mug of hot chocolate ______	**B.** 1 kg
3. a loaf of bread ______	**C.** 1 g
4. a full-grown grizzly bear ______	**D.** 50 g
5. a safety pin ______	**E.** 350 g

Solve.

6. Two paper clips weigh about 1 gram. About how many paper clips weigh 10 grams? ______

7. About how many paper clips weigh 1 kilogram? (One kilogram = 1,000 grams.) ______

8. One ounce is about 30 grams. About how many paper clips are in 1 ounce? ______

9. About how many paper clips are in 1 pound? ______

10. About how much does a box of 1,000 paper clips weigh if the empty box weighs 15 grams? ______

FACTS PRACTICE **Solve.**

11. $8 * 9 =$ ______

12. $\frac{96}{8} =$ ______

13. $60 \div 12 =$ ______

14. $12 * 7 =$ ______

15. $6 * 11 =$ ______

16. $9 * 5 =$ ______

17. $6 * 4 =$ ______

18. $\frac{80}{10} =$ ______

19. $10 * 11 =$ ______

20. $7 * 11 =$ ______

Name Date Time

Practice Set 81 *continued*

SRB 4 5 106 107

21. Complete the Powers of 10 Table.

The Powers of 10 Table

Millions	Hundred-Thousands	Ten-Thousands	Thousands	Hundreds	Tens	Ones
1,000,000				100		1
10 [100,000s]			10 [100s]			10 [0.1s]
		10∗10∗10∗10				
	10^5		10^3			10°

Multiply.

22. 4.6 ∗ 24 = ______________

23. 125 ∗ 0.01 = ______________

24. 0.61 ∗ 33 = ______________

25. 1.24 ∗ 52 = ______________

26. 0.7 ∗ 130 = ______________

27. 5.4 ∗ 361 = ______________

28. **Writing/Reasoning** Explain how you know that the pattern below is an example of a translation.

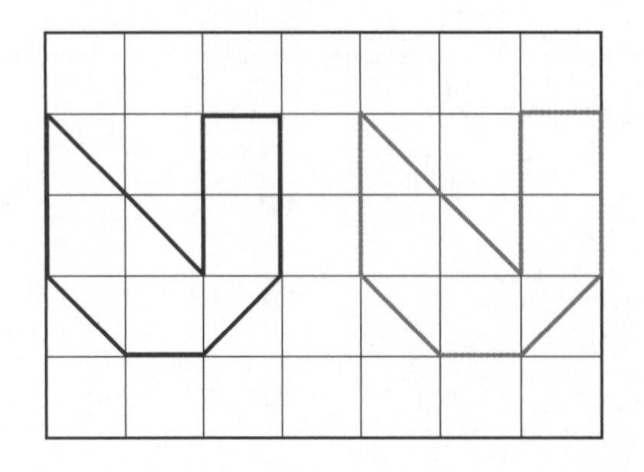

____

____

____

Name Date Time

Practice Set 82

SRB 34–37 59 92–97

Write the letter of the description that matches the polygon.

1. equilateral triangle ________

2. parallelogram ________

3. square ________

4. trapezoid ________

A. 4 sides of equal length, 4 right angles

B. no right angles, all sides the same length

C. only 1 pair of parallel sides

D. 2 pairs of parallel sides, no right angles

Answer the questions.

5. How much is $\frac{1}{8}$ of 32¢? ________ ¢

6. How much is $\frac{4}{9}$ of 54¢? ________ ¢

7. How much is $\frac{1}{10}$ of 80¢? ________ ¢

8. How much is $\frac{1}{3}$ of 90¢? ________ ¢

9. How much is $\frac{1}{5}$ of $2.20? ________ ¢

10. How much is $\frac{2}{3}$ of 27¢? ________ ¢

11. How much is $\frac{5}{7}$ of 63¢? ________ ¢

12. How much is $\frac{4}{6}$ of 66¢? ________ ¢

13. How much is $\frac{1}{3}$ of $2.01? ________ ¢

14. How much is $\frac{2}{5}$ of $2.35? ________ ¢

15. **Writing/Reasoning** Joel has a $5 bill. Explain whether he has enough money to buy 3 pens that cost $1.69 each.

____

____

____

Name Date Time

Practice Set 83

Draw the reflections.

1.

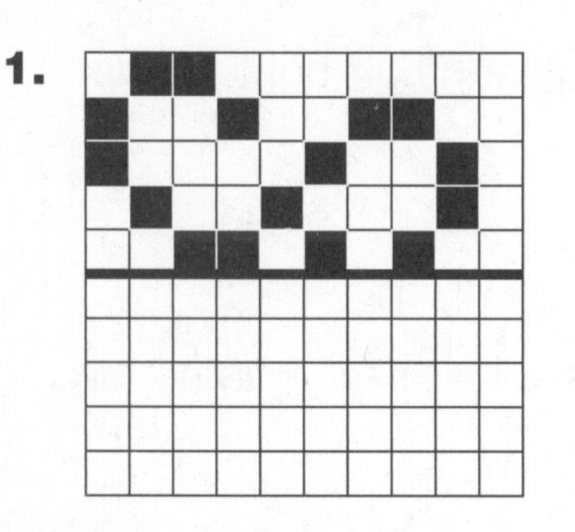

2.

Identify the rotations.

3. Which figure shows a $\frac{3}{4}$ turn counterclockwise? ________

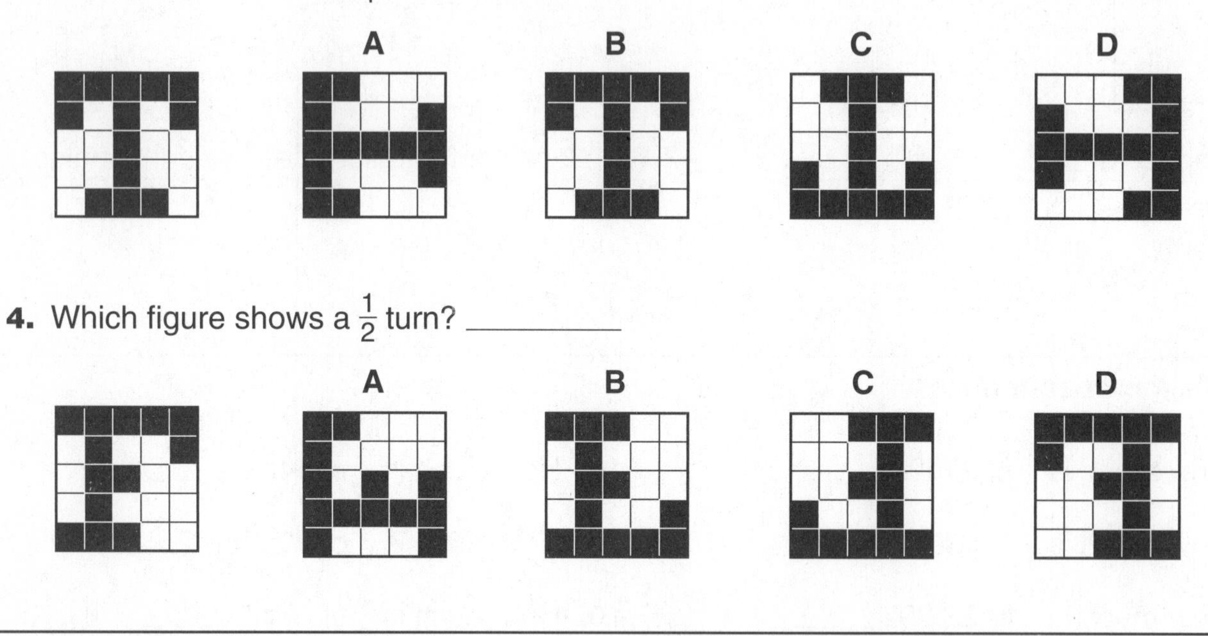

4. Which figure shows a $\frac{1}{2}$ turn? ________

5. **Writing/Reasoning** Explain why a cylinder is not a polyhedron.

____

____

____

Compare, using <, >, or =.

6. $\frac{1}{4}$ ______ $\frac{2}{3}$

7. $\frac{5}{16}$ ______ $\frac{1}{4}$

8. $\frac{5}{11}$ ______ $\frac{5}{10}$

9. 432 ______ 578

10. 17,300 ______ 13,700

11. 765,983 ______ 725,983

Name Date Time

Practice Set 84

SRB 133–136 151

Find the area of each polygon.

1. 7 in. 14 in.

Area: ______________

2. 3.2 m 3.2 m

Area: ______________

3. 9.7 cm 14 cm

Area: ______________

4. 9 ft 15 ft

Area: ______________

COMPUTATION PRACTICE **Solve.**

5. $56 + 98$

6. $13 + 73$

7. $74 - 46$

8. $30 - 19$

9. $623 + 351 + 249$

10. $403 + 382$

11. $91 * 100$

12. $5{,}348 + 6{,}155$

13. $4{,}390 - 240$

14. $(80 + 80) * 5 =$ ______________

15. $36 + (7 * 9) =$ ______________

16. $60 + 450 + 338 =$ ______________

17. $(40 * 5) - 30 =$ ______________

Name Date Time

Practice Set 84 *continued*

SRB 10–24

FACTS PRACTICE

In each problem set below, do as many problems as you can in one minute. You can ask someone to time you.

Problem Set 1

18. 14 − 6 = ________

19. 7 / 7 = ________

20. 96 / 8 = ________

21. 4 * 111 = ________

22. 11 − 4 = ________

23. 12 + 5 = ________

24. 14 + 8 = ________

25. 12 − 7 = ________

26. 54 / 9 = ________

27. 9 * 12 = ________

28. 14 / 7 = ________

29. 24 / 3 = ________

30. 5 + 6 = ________

31. 18 / 2 = ________

32. 4 * 80 = ________

Problem Set 2

33. 84 / 7 = ________

34. 12 * 11 = ________

35. 7 + 18 = ________

36. 3 * 8 = ________

37. 4 * 10 = ________

38. 11 * 4 = ________

39. 3 * 11 = ________

40. 12 + 9 = ________

41. 4 * 7 = ________

42. 11 + 11 = ________

43. 81 / 9 = ________

44. 54 / 6 = ________

45. 50 + 60 = ________

46. 40 / 10 = ________

47. 27 / 3 = ________

Problem Set 3

48. 72 / 6 = ________

49. 44 / 4 = ________

50. 121 − 11 = ________

51. 9 * 9 = ________

52. 144 / 12 = ________

53. 7 * 8 = ________

54. 4 * 70 = ________

55. 144 / 12 = ________

56. 21 − 3 = ________

57. 55 + 5 = ________

58. 6 * 7 = ________

59. 9 * 9 = ________

60. 500 + 600 = ________

61. 32 / 8 = ________

62. 36 / 9 = ________

Use with or after Lesson 11•4.

Name Date Time

Practice Set 85

SRB 49–51 137 138

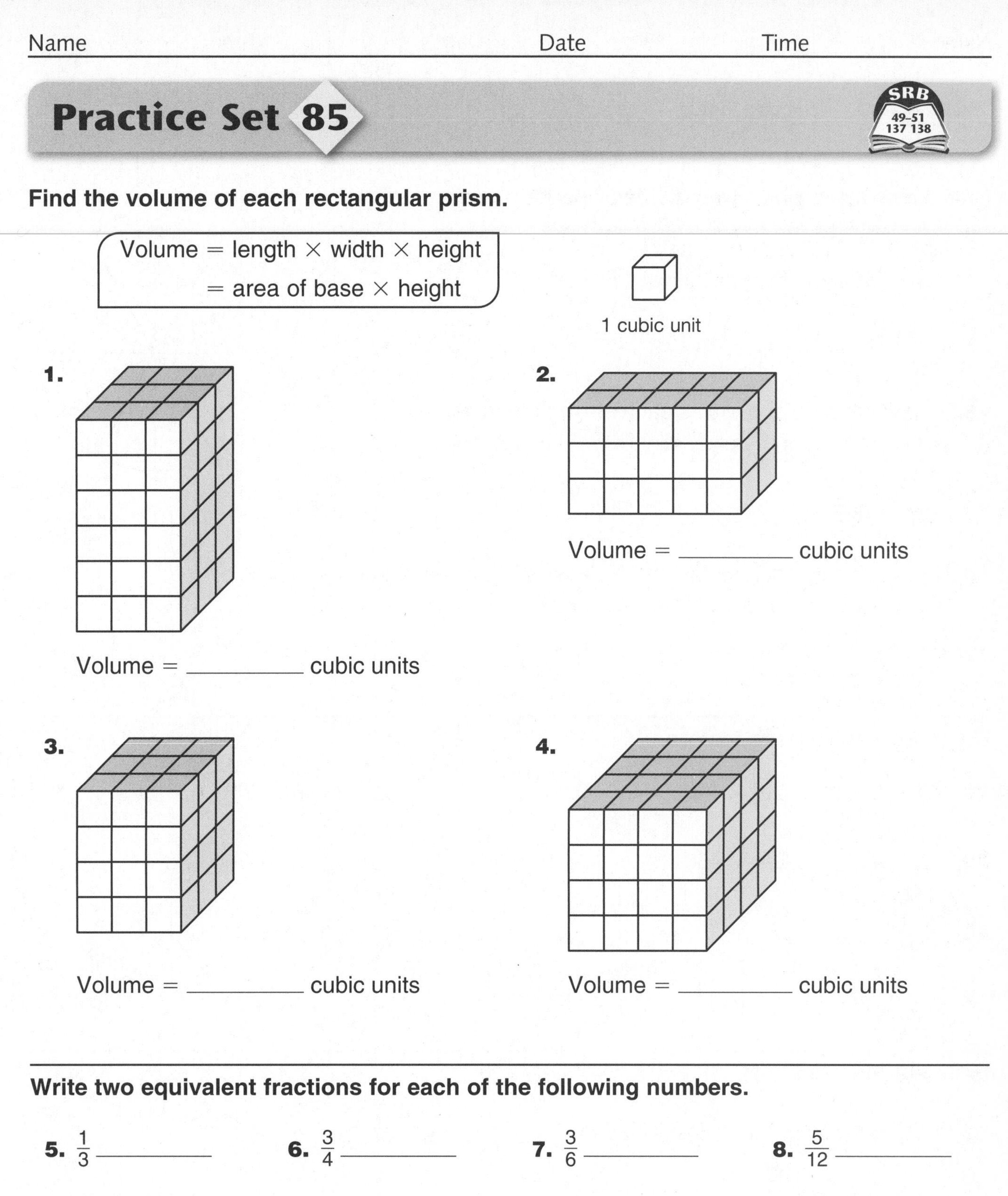

Find the volume of each rectangular prism.

Volume = length × width × height
= area of base × height

1 cubic unit

1. Volume = ________ cubic units

2. Volume = ________ cubic units

3. Volume = ________ cubic units

4. Volume = ________ cubic units

Write two equivalent fractions for each of the following numbers.

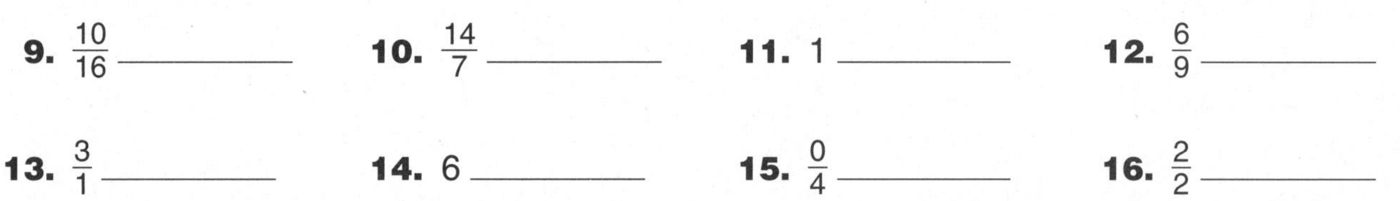

5. $\frac{1}{3}$ ________ **6.** $\frac{3}{4}$ ________ **7.** $\frac{3}{6}$ ________ **8.** $\frac{5}{12}$ ________

9. $\frac{10}{16}$ ________ **10.** $\frac{14}{7}$ ________ **11.** 1 ________ **12.** $\frac{6}{9}$ ________

13. $\frac{3}{1}$ ________ **14.** 6 ________ **15.** $\frac{0}{4}$ ________ **16.** $\frac{2}{2}$ ________

Name Date Time

Practice Set 85 *continued*

SRB 7 44 45 175

One super-sized pizza was divided into 12 pieces.

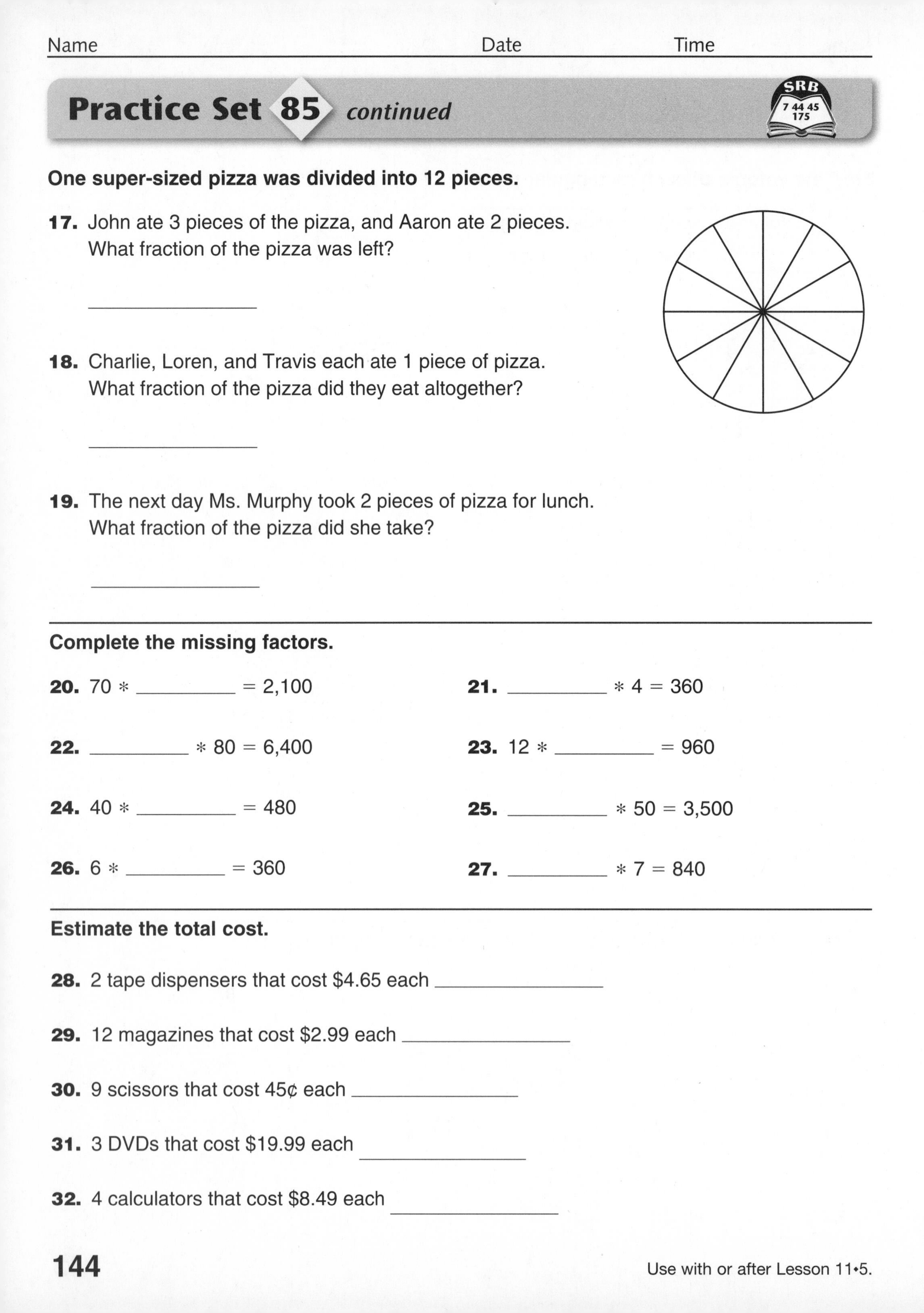

17. John ate 3 pieces of the pizza, and Aaron ate 2 pieces. What fraction of the pizza was left?

____

18. Charlie, Loren, and Travis each ate 1 piece of pizza. What fraction of the pizza did they eat altogether?

____

19. The next day Ms. Murphy took 2 pieces of pizza for lunch. What fraction of the pizza did she take?

____

Complete the missing factors.

20. 70 * __________ = 2,100

21. __________ * 4 = 360

22. __________ * 80 = 6,400

23. 12 * __________ = 960

24. 40 * __________ = 480

25. __________ * 50 = 3,500

26. 6 * __________ = 360

27. __________ * 7 = 840

Estimate the total cost.

28. 2 tape dispensers that cost $4.65 each ______________

29. 12 magazines that cost $2.99 each ______________

30. 9 scissors that cost 45¢ each ______________

31. 3 DVDs that cost $19.99 each ______________

32. 4 calculators that cost $8.49 each ______________

Name Date Time

Practice Set 86

SRB 16 20 32 53 54 60

COMPUTATION PRACTICE

Add or subtract.

1. $6 - (-10) =$ ______

2. $5 + (-1) =$ ______

3. $-6 + 6 =$ ______

4. $17 - (-2) =$ ______

5. $-11 - (-15) =$ ______

6. $16 - (-7) =$ ______

7. $-19 + (-6) =$ ______

8. $-21 - 9 =$ ______

Write the numbers from least to greatest.

9. $2.3, -7, \frac{4}{9}, -1.5, 8.3, -0.2$

______, ______, ______, ______, ______, ______

10. $-11, 3\frac{2}{5}, 1.85, -5.5, 4, -\frac{8}{10}$

______, ______, ______, ______, ______, ______

11. $\frac{17}{10}, -4, 9.9, 9.09, -3.7, 1.07$

______, ______, ______, ______, ______, ______

COMPUTATION PRACTICE

Solve.

12. $\frac{540}{6} =$ ______

13. $250 * 80 =$ ______

14. $640 = 8 *$ ______

15. $20 * 300 =$ ______

16. $60 *$ ______ $= 2,400$

17. ______ $\div 50 = 6$

18. $\frac{5,600}{700} =$ ______

19. $360 *$ ______ $= 7,200$

20. ______ $\div 5 = 35$

21. $9 * 200 =$ ______

22. $540 \div$ ______ $= 90$

23. $110 * 120 =$ ______

Name ______________________ Date ____________ Time ____________

Practice Set 86 *continued*

SRB 61 62 93 129

Rename the following numbers as percents.

24. $\frac{1}{4}$ = ______% **25.** 0.75 = ______% **26.** 1.00 = ______% **27.** $\frac{57}{100}$ = ______%

28. $\frac{3}{20}$ = ______% **29.** $\frac{10}{25}$ = ______% **30.** 0.4 = ______% **31.** $\frac{37.5}{100}$ = ______%

32. $\frac{4}{5}$ = ______% **33.** 1.125 = ______% **34.** $\frac{765}{1,000}$ = ______% **35.** $\frac{6}{15}$ = ______%

---

1 km = 1000 m; 1 m = 100 cm
1 cm = 10 mm

Complete.

36. 2 km = __________ cm **37.** 25,000 mm = __________ m

38. 1,800 m = __________ km **39.** 30 km = __________ m

40. 3.3 cm = __________ mm **41.** 670 cm = __________ mm

Estimate the angle measures.

42. __________________

43.

____

44. __________________

45. __________________

Use with or after Lesson 11•6.

Name Date Time

Practice Set 87

SRB 12–15 18 137 156 180

Complete.

1. 4 cups = __________ pints

2. 3 quarts = __________ cups

3. __________ gallons = 8 quarts

4. __________ pints = 10 cups

5. $1\frac{1}{2}$ quarts = __________ cups

6. $3\frac{1}{2}$ gallons = __________ quarts

---

7. What time does the clock show? Write your answer to the nearest minute.

____

8. What time will it be in 35 minutes?

____

9. What time will it be in 88 minutes?

____

Find the percentages.

10. 70% of 10 __________

11. 25% of 80 __________

12. 75% of 12 __________

13. 50% of 64 __________

14. 24% of 25 __________

15. 150% of 22 __________

16. 80% of 50 __________

17. 90% of 100 __________

18. 33% of 1,000 __________

19. 12% of 200 __________

20. 6% of 50 __________

21. 15% of 20 __________

Solve these problems mentally.

22. 934,167 − 1,000 = __________

23. 934,167 − 100 = __________

24. 934,167 − 10,000 = __________

25. 934,167 − 10 = __________

Multiply.

26. 2 * 538 = __________

27. 1,207 * 5 = __________

28. 8 * 120 = __________

Name Date Time

Practice Set 88

Write the missing numbers.

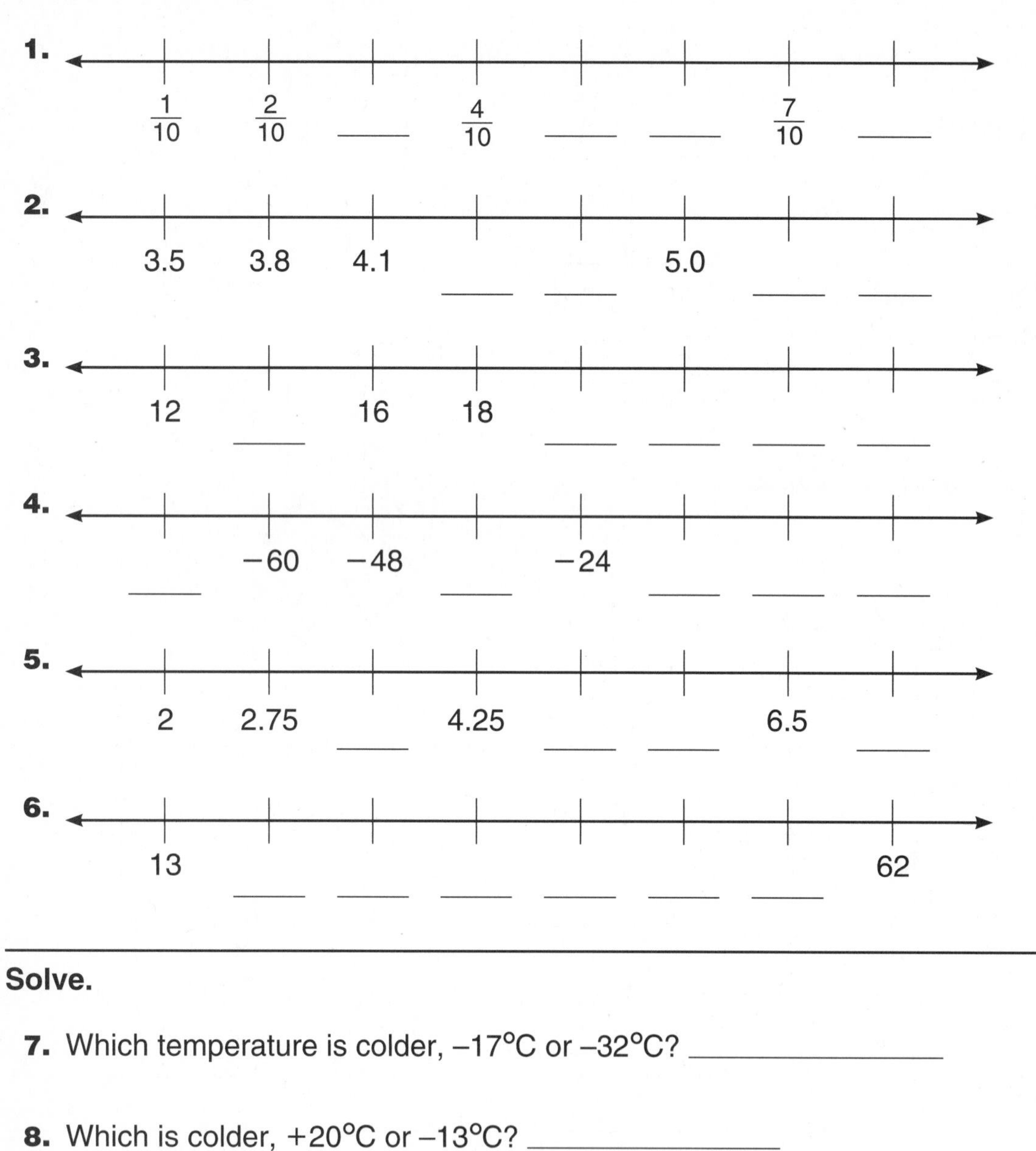

Solve.

7. Which temperature is colder, –17°C or –32°C? ________________

8. Which is colder, +20°C or –13°C? ________________

9. You have 62 cookies. You give 4 cookies to each of your friends and you eat whatever is left. How many friends received cookies? ________________

10. You have 62 cookies, and you want to share them equally among you and 17 friends. How many will each person get? ________________

Name Date Time

Practice Set 89

Complete the rate tables for the problems below.
Then answer the questions.

Example Richard's car travels about 25 miles on 1 gallon of gasoline.

miles	25	50	75	100	125	150
gallons	1	2	3	4	5	6

1. How far can the car travel on 6 gallons of gas? ______________

2. At 125 miles, how many gallons have been used? ______________

The Perfect Paint factory can make 100 gallons of paint per day.

gallons	100						
day	1	2	3	4	5	6	7

3. How many gallons can the factory make in a week? ______________

4. How many gallons can the factory make in a year? ______________

Rita is a seamstress. She can make 3 dresses in 2 hours.

dresses	3					
hours	2	4	6	8	10	12

5. How many dresses can Rita make in an 8-hour day? ______________

6. If she takes off an hour for lunch, how many dresses will she make? ______________

Write <, >, or = to make a true number sentence.

7. 29 + 33 __________ 45 + 17

8. 20 + 30 + 40 __________ 35 + 55

9. 35 + 13 __________ 32 + 18

10. 130 − 18 __________ 55 ∗ 2

11. 13 + 12 __________ 48 ÷ 2

12. 118 + 220 __________ 1,000 − 692

Name Date Time

Practice Set 89 *continued*

SRB 5 38–39 150 155

Find the percentage.

13. 34% of 50 ________

14. 25% of 60 ________

15. 15% of 100 ________

16. 60% of 600 ________

17. 1% of 20 ________

18. 10% of 77 ________

19. 5% of 20 ________

20. 66% of 1,000 ________

21. 22% of 150 ________

22. 45% of 900 ________

23. 10% of 300 ________

24. 100% of 93 ________

Solve each number sentence by finding the value of the variable.

25. $a = (8 * 9) \div 3$ ________

26. $\frac{64}{b} = 32 \div 2$ ________

27. $c = (23 + 9) \div 8$ ________

28. $45 + (11 * 6) = d$ ________

29. $43 - e = \frac{80}{2}$ ________

30. $\frac{f}{19} = 570$ ________

31. $(60 - 18) \div 6 = g$ ________

32. $(13 * 20) - h = 10^2$ ________

Complete.

33. $10^4 =$ ________

34. $10^{\square} = 100{,}000$

35. $600 = 6 * 10^{\square}$

36. $10 * 10 * 10 * 10 = 10^{\square}$

37. 10 to the eighth power = ________

38. $3.0 * 10^6 =$ ________

39. 10 to the ________ power = 1,000

40. $10^0 =$ ________

Name Date Time

Practice Set 90

SRB 47 178

Complete the rate tables for the problems below. Then answer the questions.

It takes Christie 3 minutes to read a page of her book.

minutes	3					
pages	1	2	3	4	5	6

1. How many pages can she read in 15 minutes? ____________
2. At this rate, how many pages will she read in $\frac{1}{2}$ hour? ____________

The cars on the freeway are traveling 55 miles per hour.

miles	55					
hours	1	2	3	4	5	6

3. How far will they go in 5 hours? ____________
4. About how long will it take to travel 300 miles? ____________

Howard delivers 14 newspapers in 10 minutes.

papers	14					
minutes	10	20	30	40	50	60

5. How many papers can Howard deliver in 1 hour? ____________
6. How long will it take Howard to deliver the 70 papers on his route? ____________

Tomatoes cost 65¢ a pound.

price	$1.30					
pounds	2	4	6	8	10	12

7. How much do 10 pounds of tomatoes cost? ____________
8. About how many pounds of tomatoes can you buy with $5.00? ____________

Name Date Time

Practice Set 90 *continued*

SRB 5 61 137 138

Rename the following fractions as decimals.

9. $\frac{8}{100}$ ______ **10.** $\frac{6}{10}$ ______ **11.** $\frac{3}{10}$ ______ **12.** $\frac{1}{3}$ ______

13. $\frac{250}{1,000}$ ______ **14.** $\frac{47}{100}$ ______ **15.** $\frac{4}{8}$ ______ **16.** $\frac{1}{4}$ ______

17. $\frac{6}{16}$ ______ **18.** $\frac{42}{100}$ ______ **19.** $\frac{182}{1,000}$ ______ **20.** $\frac{14}{18}$ ______

Write the next three numbers in the pattern.

21. 23,610; 23,615; 23,620; ______; ______; ______

22. 39.55, 39.50, 39.45, ______, ______, ______

23. 151, 148, 145, ______, ______, ______

24. 1,455; 1,130; 805; ______; ______; ______

Complete.

25. $10^2 =$ ______

26. $10^{\square} = 1,000$

27. $10 * 10 * 10 * 10 * 10 * 10 = 10^{\square}$

28. $10^{\square} = 10,000$

Volume = length × width × height
= area of base × height

Find the volume of each rectangular prism.

29.

______ cubic units

30.

______ cubic units

Use with or after Lesson 12•3.

Name Date Time

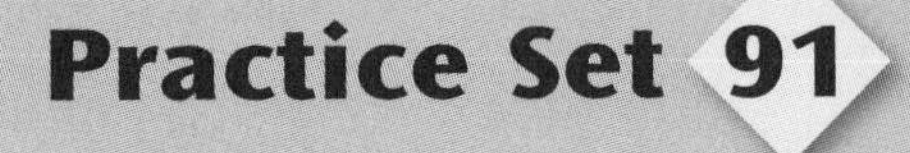

SRB
47 59
175–178

Solve.

1. Mr. Turner drives 260 miles and uses 10 gallons of gas. How far could he drive on 1 gallon of gas?

____

2. Ms. Smith buys 3 pounds of apples for $2.04. What is the price for 1 pound of apples?

____

3. Allison sells 6 cards for $9.00. What is the price per card?

____

4. Tonya jogs 8 miles in 1 hour 12 minutes. What is her rate per mile?

____

Answer the following questions. If the answer is a fraction, write it in simplest terms.

5. What fraction of the group of coins is pennies? _______________

6. What fraction of the group of coins is nickels? _______________

7. What fraction of the group of coins is dimes? _______________

8. What is the value of all the coins? _______________

9. If you took away $\frac{2}{3}$ of the nickels and $\frac{5}{6}$ of the pennies, how much money would be left? _______________

Name Date Time

Practice Set 92

SRB 6 47 53 54 149

Find the unit price for each. Tell which is the better buy.

1. **a.** 4 cans of peaches for $1.00 ________________

b. 2 cans of peaches for $0.60 ________________

Better buy: ________________

2. **a.** 6 ounces of raisins for $1.68 ________________

b. 1 pound of raisins for $3.52 ________________

Better buy: ________________

3. **a.** 8 eggs for $2.00 ________________

b. 1 dozen eggs for $2.40 ________________

Better buy: ________________

4. **a.** 3 juice boxes for $1.05 ________________

b. 10 juice boxes for $3.50 ________________

Better buy: ________________

Write <, >, or = to make each number sentence true.

5. $47 + 63$ __________ $22 + 74$

6. $8 + 43$ __________ $7 + 13 + 31$

7. $85 + 23$ __________ $81 + 35$

8. $9 * 12$ __________ $\frac{404}{4}$

9. $169 - 40$ __________ $95 + 26$

Order these numbers from least to greatest.

10. $\frac{1}{4}$ $\frac{3}{6}$ $\frac{1}{10}$ $\frac{9}{12}$ $\frac{16}{16}$

______, ______, ______, ______, ______

Use with or after Lesson 12•5.

Name Date Time

Practice Set 92 *continued*

SRB 44 45 80–84 149

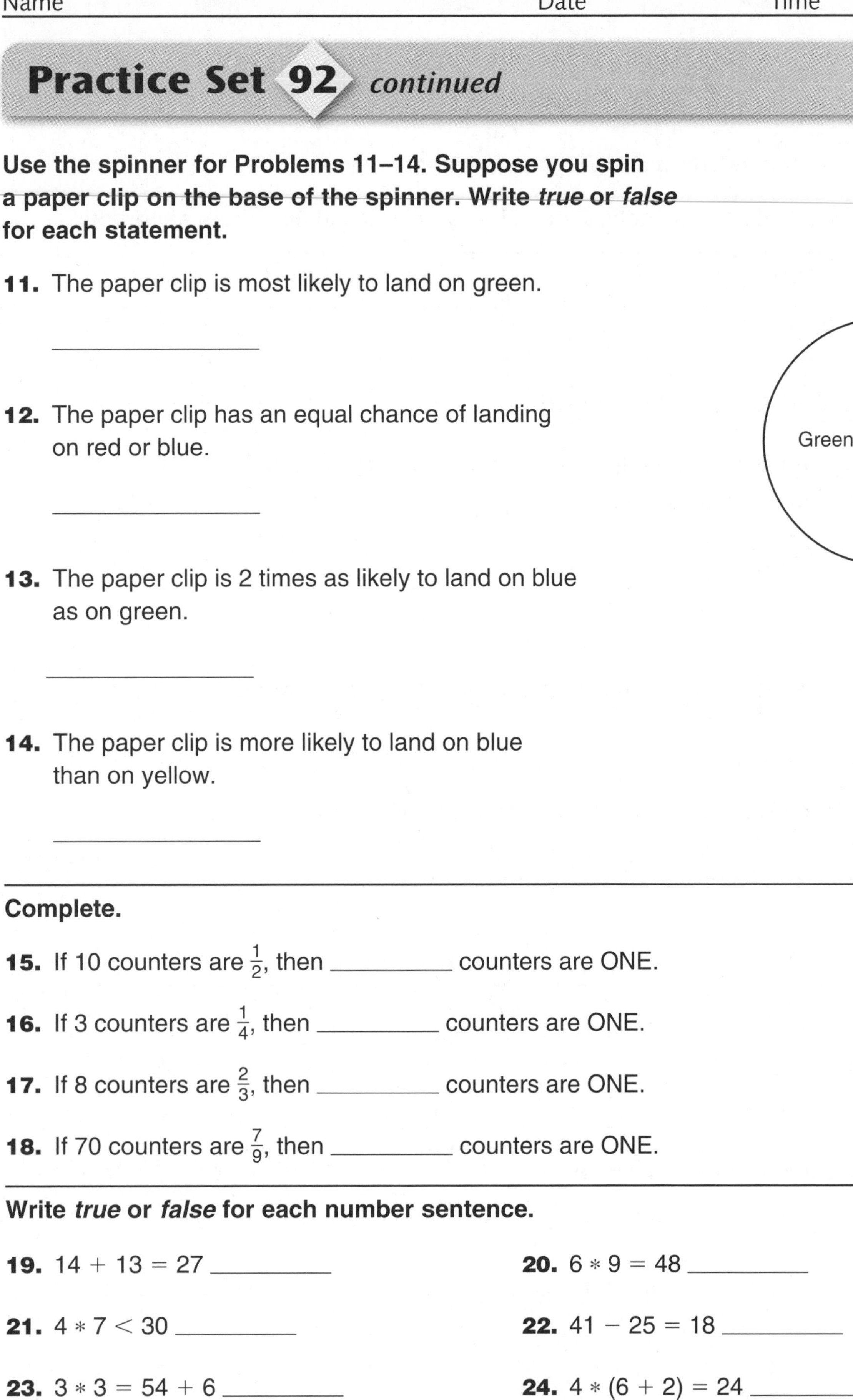

Use the spinner for Problems 11–14. Suppose you spin a paper clip on the base of the spinner. Write *true* or *false* for each statement.

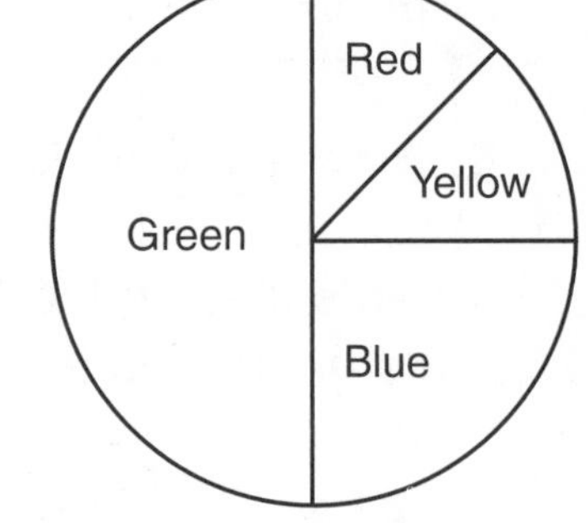

11. The paper clip is most likely to land on green.

____

12. The paper clip has an equal chance of landing on red or blue.

____

13. The paper clip is 2 times as likely to land on blue as on green.

____

14. The paper clip is more likely to land on blue than on yellow.

____

Complete.

15. If 10 counters are $\frac{1}{2}$, then ________ counters are ONE.

16. If 3 counters are $\frac{1}{4}$, then ________ counters are ONE.

17. If 8 counters are $\frac{2}{3}$, then ________ counters are ONE.

18. If 70 counters are $\frac{7}{9}$, then ________ counters are ONE.

Write *true* or *false* for each number sentence.

19. $14 + 13 = 27$ ________

20. $6 * 9 = 48$ ________

21. $4 * 7 < 30$ ________

22. $41 - 25 = 18$ ________

23. $3 * 3 = 54 + 6$ ________

24. $4 * (6 + 2) = 24$ ________

Name Date Time

Practice Set 93

Complete the rate tables. Then answer the questions.

James is selling tickets for the school play. He is able to sell 60 tickets each hour.

Tickets	60					
Hours	1	2	3	4	5	6

1. How many hours will it take to sell 360 tickets? ______________ hours

Ms. Chen's class collects soup labels for their school. Each of the 24 students brings in 5 labels each week.

Labels	120					
Weeks	1	2	3	4	5	6

2. How many soup labels will the class have at the end of six weeks?

______________ soup labels

Mr. Patel is driving on the highway. He has traveled 65 miles after the first hour and continues driving at this pace.

Miles	65					
Hours	1	2	3	4	5	6

3. How many miles will Mr. Patel have driven after 4 hours?

______________ miles

Round each decimal to the nearest tenth.

4. 32.79 ______________

5. 8.43 ______________

6. 674.12 ______________

7. 17.98 ______________

8. 1.65 ______________

9. 1,905.437 ______________

Use with or after Lesson 12•6.

Name Date Time

Test Practice 1

Fill in the circle next to your answer.

1. Tennessee covers an area of 42,146 square miles. What is the place value of the digit 1 in 42,146?

Ⓐ tens Ⓑ hundreds Ⓒ thousands Ⓓ ten thousands

2. The chart below shows Kiki's vocabulary quiz scores. What is Kiki's **median** quiz score?

Kiki's Vocabulary Quiz Scores

Quiz 1	Quiz 2	Quiz 3	Quiz 4	Quiz 5
26	21	21	22	25

Ⓐ 4 Ⓑ 21 Ⓒ 22 Ⓓ 23

3. This line graph shows the daily low temperatures in Jacksonville, Florida, for five days.

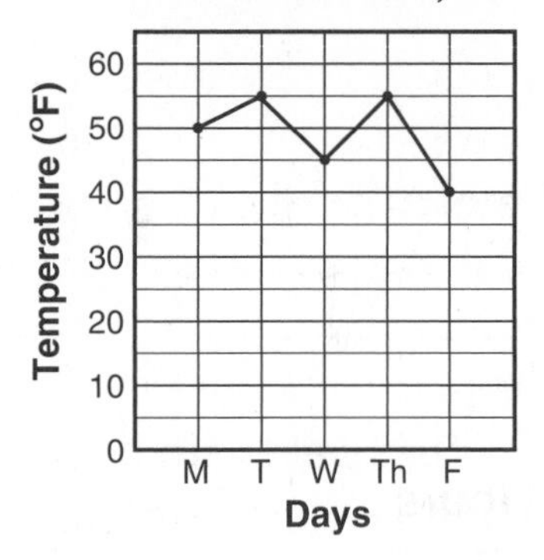

What is the **range** of the temperatures shown on this graph?

Ⓐ 15° Ⓑ 49° Ⓒ 50° Ⓓ 55°

4. Look at the circle below.

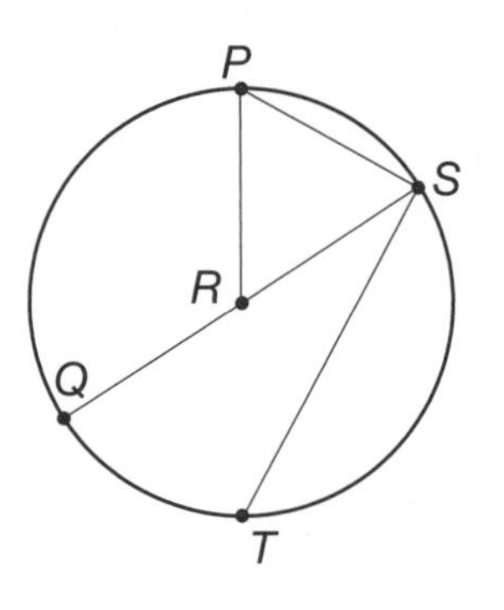

Which of the following is a **radius** of the circle?

Ⓐ $\overline{PS}$ Ⓑ $\overline{ST}$ Ⓒ $\overline{RP}$ Ⓓ $\overline{QS}$

Test Practice 1 *continued*

5. Look at the multiplication sentences below.

$$10 \times 76 = 760$$
$$100 \times 76 = 7{,}600$$
$$1{,}000 \times 76 = \square$$

What is the missing product?

Ⓐ 76,000 Ⓑ 7,600 Ⓒ 760 Ⓓ 76

6. Greg used a protractor to draw a right angle. Which of the following shows the angle Greg drew?

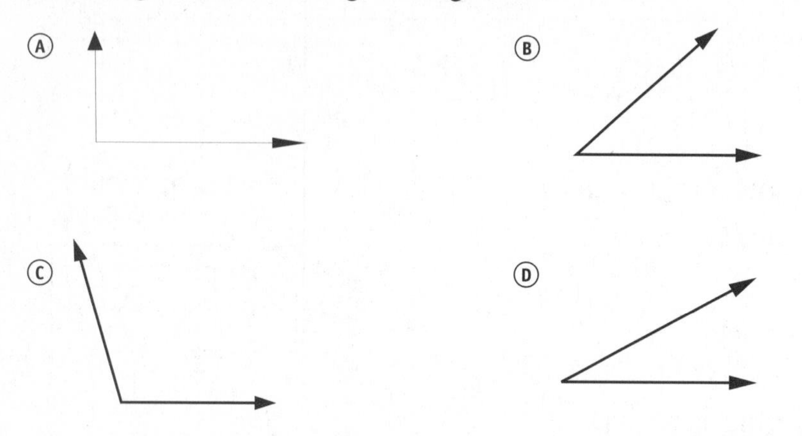

7. In 1787, Delaware became a state. Florida became a state in 1845. Which expression should you use to find out how many years passed between the time Delaware became a state and Florida became a state?

Ⓐ $1845 + 1787$ Ⓑ $1845 - 1787$

Ⓒ 1845×1787 Ⓓ $1845 \div 1787$

8. Which pair of circles is **concentric**?

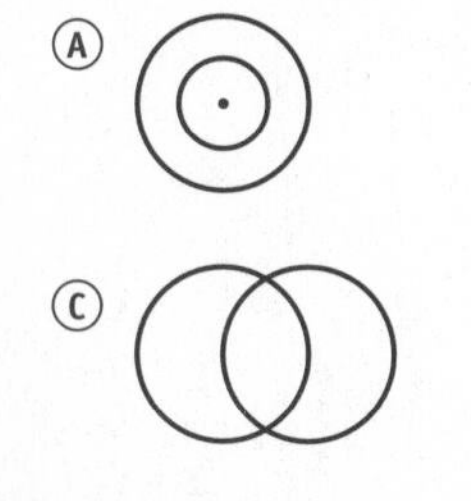

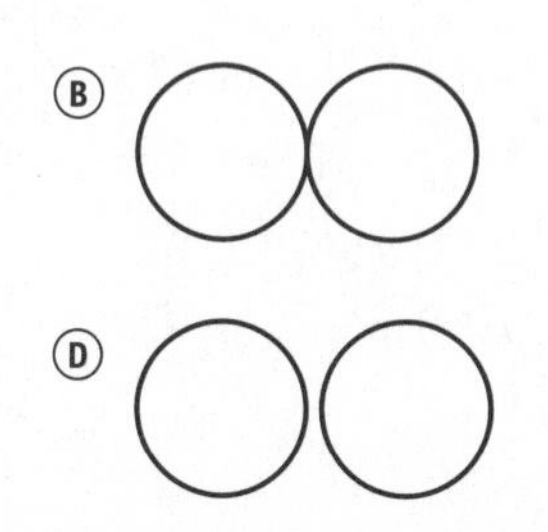

Name Date Time

Test Practice 2

Fill in the circle next to your answer.

1. A black-chinned hummingbird is 8.75 centimeters long. A blue-throated hummingbird is 12.8 centimeters long. How much longer is the blue-throated hummingbird than the black-chinned hummingbird?

Ⓐ 3.05 centimeters Ⓑ 4.05 centimeters

Ⓒ 4.15 centimeters Ⓓ 21.55 centimeters

2. In Mr. Johnson's fourth grade class, $\frac{11}{20}$ of the students choose art as their favorite subject. Which of the following has the same value as $\frac{11}{20}$?

Ⓐ 0.11 Ⓑ 0.22 Ⓒ 0.44 Ⓓ 0.55

3. Mandy uses these blocks to show the decimal 0.324.

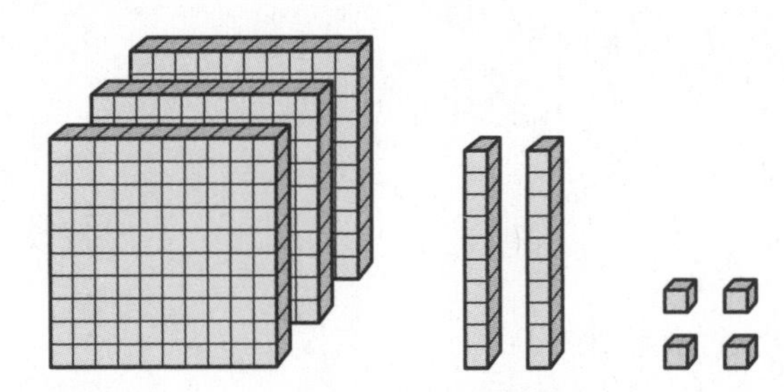

Mandy wants to subtract 0.05 from this number. How can she use the blocks to help her subtract?

Ⓐ Trade 1 long for 10 cubes, and then subtract 5 cubes.

Ⓑ Trade 1 flat for 10 cubes, and then subtract 5 cubes.

Ⓒ Trade 1 long for 10 flats, and then subtract 5 longs.

Ⓓ Trade 1 flat for 10 longs, and then subtract 5 longs.

Name Date Time

Test Practice 2 *continued*

4. A stadium has 378 seats. Each week, 21 games are played in the stadium. The owner estimates that about 8,000 people can see a game each week. Which of the following correctly tells how the owner made her estimate?

Ⓐ She multiplied 20×300.

Ⓑ She multiplied 20×400.

Ⓒ She multiplied 30×300.

Ⓓ She multiplied 30×400.

5. Caroline drew this picture.

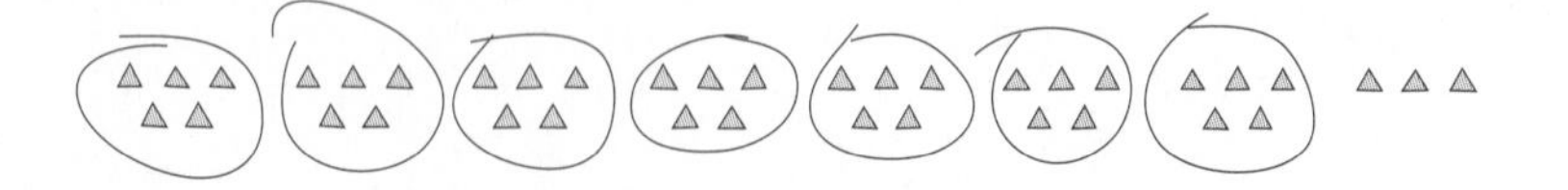

Which of the following does the picture represent?

Ⓐ $38 \div 7 = 5$

Ⓑ $38 \div 7 = 5$ R3

Ⓒ $35 \div 7 = 5$ R3

Ⓓ $35 \div 7 = 5$

6. The jar below contains peanuts. What is the **best** estimate of the total number of peanuts in the jar?

Ⓐ about 2,000 peanuts

Ⓑ about 800 peanuts

Ⓒ about 400 peanuts

Ⓓ about 200 peanuts

Name Date Time

Test Practice 3

Fill in the circle next to your answer.

1. The circle graph below shows how Derek spends his allowance each month.

DEREK'S MONTHLY ALLOWANCE

Movies, Savings, Video Games, Magazines

On what does Derek spend $\frac{1}{8}$ of his money?

Ⓐ movies Ⓑ savings

Ⓒ video games Ⓓ magazines

2. In a game of darts, Conrad hit 8 bull's-eyes in 25 shots. Which of the following has the same value as $\frac{8}{25}$?

Ⓐ 0.8 Ⓑ 0.16

Ⓒ 0.25 Ⓓ 0.32

3. A tree swallow is 5 inches long. A mallard duck is 25 inches long. The length of the tree swallow is $\frac{5}{25}$ of the mallard's length.

Which fraction has a value equal to $\frac{5}{25}$?

Ⓐ $\frac{1}{5}$ Ⓑ $\frac{1}{4}$

Ⓒ $\frac{1}{2}$ Ⓓ $\frac{25}{2}$

Name Date Time

Test Practice 3 *continued*

Fill in the circle next to your answer.

4. Cory randomly places the cards below facedown on a table.

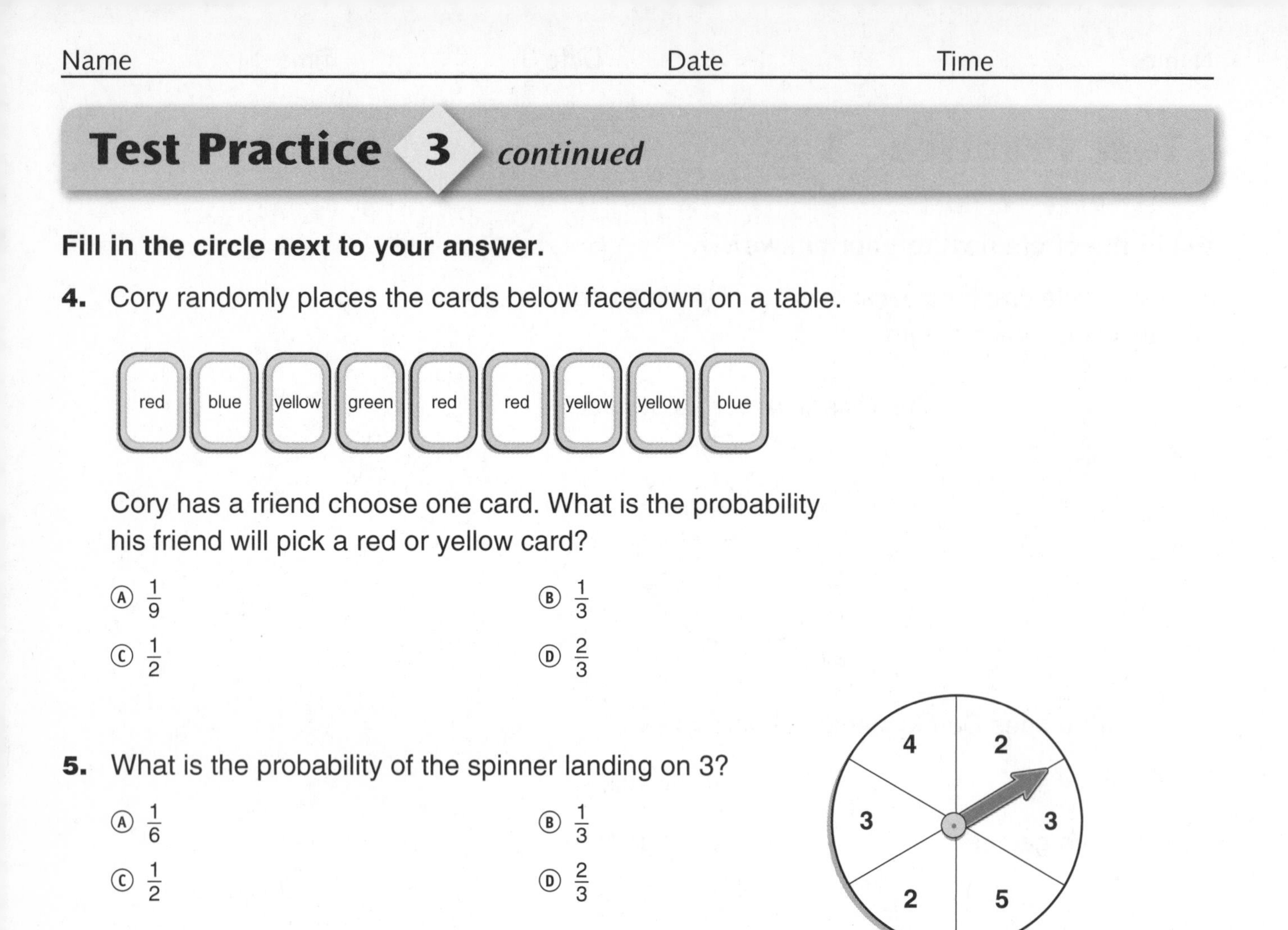

Cory has a friend choose one card. What is the probability his friend will pick a red or yellow card?

Ⓐ $\frac{1}{9}$ Ⓑ $\frac{1}{3}$

Ⓒ $\frac{1}{2}$ Ⓓ $\frac{2}{3}$

5. What is the probability of the spinner landing on 3?

Ⓐ $\frac{1}{6}$ Ⓑ $\frac{1}{3}$

Ⓒ $\frac{1}{2}$ Ⓓ $\frac{2}{3}$

6. Ms. Green's storeroom is represented by the shaded figure on the grid.

What is the perimeter of her storeroom?

Ⓐ 30 feet Ⓑ 28 feet

Ⓒ 25 feet Ⓓ 22 feet

□ = 1 square foot

7. Ms. Alberti is buying carpet for her study. The picture shows the size and shape of her study.

How much carpet does Ms. Alberti need?

Area of a rectangle = length × width

Ⓐ 34 square feet Ⓑ 68 square feet

Ⓒ 280 square feet Ⓓ 400 square feet

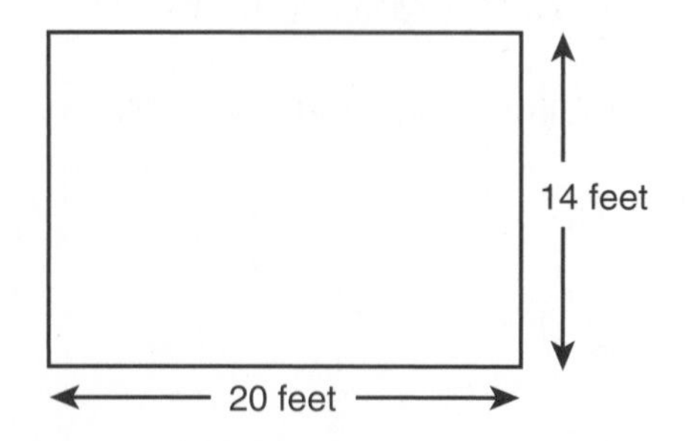

Name Date Time

Test Practice 4

Fill in the circle next to your answer.

1. Which pair of coordinates identifies point *M*?

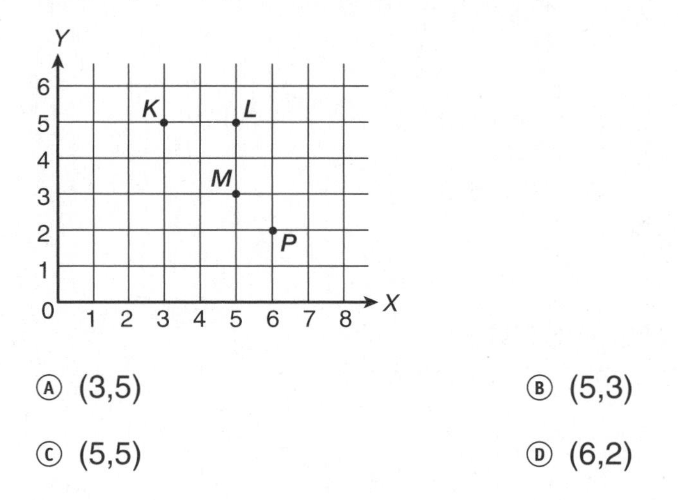

Ⓐ (3,5)

Ⓑ (5,3)

Ⓒ (5,5)

Ⓓ (6,2)

2. Which shows the figure after it is flipped over the dotted line?

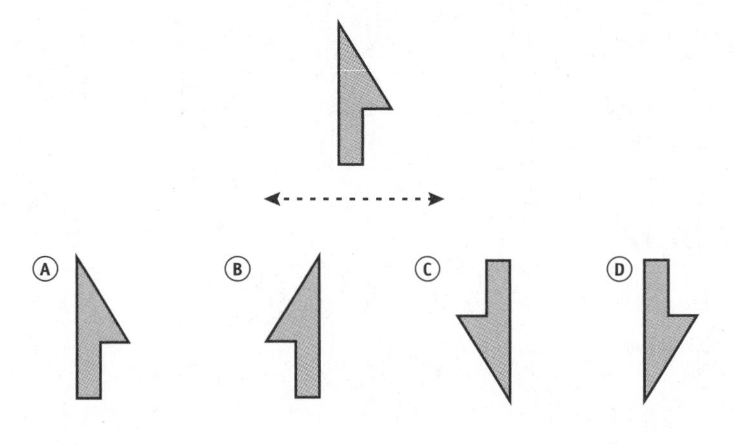

3. What movement is shown in the picture below?

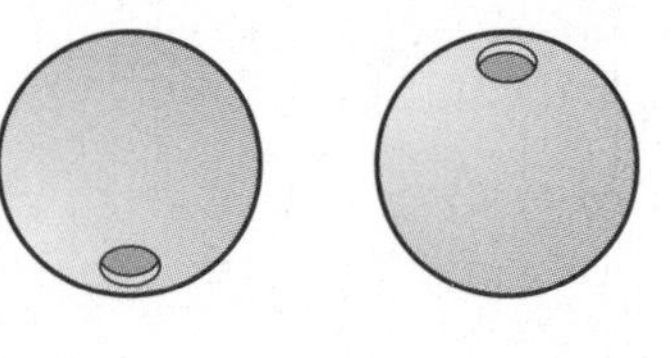

Ⓐ a rotation of 90°

Ⓑ a rotation of 180°

Ⓒ a rotation of 360°

Ⓓ a slide along a line

Name Date Time

Test Practice 4 *continued*

Fill in the circle next to your answer.

4. Which operations can be used in the boxes below to get the **least** possible result?

12 □ 3 □ 3

Ⓐ + and ÷ Ⓑ ÷ and −

Ⓒ ÷ and + Ⓓ − and +

5. Tim has twice as many baseball cards as Monica. Let *c* stand for the number of baseball cards Monica has. Which of the following represents the number of cards Tim has?

Ⓐ $2 + c$ Ⓑ $c - 2$

Ⓒ $c \times 2$ Ⓓ $c \div 2$

6. Nam's older brother earned $161 for 7 hours of work. How much money did his brother earn per hour?

Ⓐ $20 Ⓑ $22

Ⓒ $23 Ⓓ $1,127

7. Which of these number sentences is true?

Ⓐ $13 < 11$ Ⓑ $4 < -7$

Ⓒ $-6 < -5$ Ⓓ $-15 > -12$

8. Which number shows 0.726 rounded to the hundredths place?

Ⓐ 1.00 Ⓑ 0.70

Ⓒ 0.72 Ⓓ 0.73

9. What is the product of 1.30 and 0.2?

Ⓐ 0.026 Ⓑ 0.26

Ⓒ 2.60 Ⓓ 26.0

Use with or after Unit 12.